# Dr. Roland E. Best

*Assistant Vice President, Engineering*
*Sandoz AG      Basel, Switzerland*

# PHASE-LOCKED LOOPS
## Theory, Design, and Applications

## McGRAW-HILL BOOK COMPANY

*New York   St. Louis   San Francisco   Auckland   Bogotá   Hamburg   Johannesburg*
*London   Madrid   Mexico   Montreal   New Delhi   Panama   Paris   São Paulo*
*Singapore   Sydney   Tokyo   Toronto*

Library of Congress Cataloging in Publication Data

Best, Roland E.
  Phase-locked loops.

  Bibliography: p.
  Includes index.
  1. Phase-locked loops.   I. Title.
TK7872.P38B48 1984        621.3815′33         83-18686
ISBN 0-07-005050-3

1 2 3 4 5 6 7 8 9 0   DOC DOC   8 9 8 7 6 5 4 3

ISBN 0-07-005050-3

The editors for this book were Harry L. Helms and Ruth L.
Weine, the designer was Elliot Epstein, and the production
supervisor was Sally Fliess. It was set in Century Schoolbook by
University Graphics, Inc.

Printed and bound by R. R. Donnelley & Sons, Inc.

# CONTENTS

TK
7872
.P38
B48
1984

# *PREFACE*

The phase-locked loop was introduced in 1932 by de Bellescize.* Considered an exotic device in those days, it gained increased interest in the mid-sixties when it first became available as an integrated circuit.

The phase-locked loop is found today in every home—in television receivers, stereo radios, or CB equipment—so you wouldn't expect a helpless reaction from an electronic engineer after you suggested consideration of a PLL for solving a particular problem. The engineer's perplexed question, "What the hell is a PLL?" may have been answered 100 years ago by the German humorist Wilhelm Busch in his story about the peasant and the fox.

*H. de Bellescize, "La reception synchrone," *Onde Electrique,* vol. 11, June 1932, pp. 230–240.

| The fox, though tied, is very sly—<br>Slips quickly through the hole nearby<br>And sits without; Jack's in, as seen.<br>The garden wall is in between. | Jack mounts the fence, aims his hatchet<br>For one huge blow—Sir Fox will catch it! | Poor Jack! His hatchet missed the mark<br>The scene's reversed—oh, what a lark!<br>For Jack's now out—fox in, as seen,<br>And the wall is still between. |

From Wilhelm Busch, *Der Fuchs (The Fox)*, Rascher-Verlag, Zurich, Switzerland

Killing a fox with a hatchet has been a difficult task at any time, particularly when there was a *phase error* of as much as 180° between hunter and fox. As the drawings show, the peasant tried to *capture* the fox, but was unable to get *synchronized*. Apparently both fox and peasant crossed the wall with exactly the same frequency, but the peasant wasn't able to obtain *phase tracking*. So the story ended tragically; the hatchet hit not the fox, but the rope that held it, and the fox escaped with the stolen chicken.

If a phase-locked loop locks out, our electronic engineer loses no more than the signal for the moment, but if the unlocked condition persists, the engineer's job may be lost. This book was written to reduce the risk of such an inconvenience.

Chapter 1 is so short that it can be read even by engineering managers. It tells, in a few words, what a PLL is and what it does. Chapter 2 presents a classification of the different types of PLLs. It is shown that the PLL was originally an *analog* workpiece. However, it has been drifting into *digital* territory slowly but steadily for a while, so the *software-based* PLL is no more exotic today then most other microprocessor applications.

Chapter 3 is a review of the theory of the *linear* PLL. Among other matters, the performance of the PLL in the presence of noise is given particular consideration.

In Chapter 4, the theory of the *digital* PLL is presented. Hardware and software implementations of the digital PLL are considered as well.

Chapters 5 to 8 deal with practical aspects of the PLL. Chapter 5 discusses a procedure of *designing* a practical PLL circuit. It lists a number of criteria on which the key decisions in designing a PLL should be based (e.g., linear or digital PLL, type of phase detector, or type of loop filter).

Chapter 6 offers a table listing the presently available PLL ICs including related circuits. Many practical hints on adding external elements or circuits are given.

Chapter 7 describes a large variety of PLL applications, including many design examples. A number of practical circuits are worked out step by step. Experimental results of many practical circuits are also listed.

Chapter 8 covers measuring techniques, a topic generously neglected in the field of PLL literature. It shows how to measure the relevant parameters of the PLL with typical off-the-shelf laboratory instruments such as a scope and a simple waveform generator.

A number of Appendixes analyze the dynamic performance of linear and digital PLLs in more detail. Appendix F is an introduction to the Laplace transform specially written for engineers. The author suggests that engineers not familiar with the techniques of the Laplace transform use this Appendix as an introductory lecture.

# 1 OPERATING PRINCIPLES OF THE PLL

The phase-locked loop (PLL) helps keep parts of our world orderly. If we turn on a television set, a PLL will keep heads at the top of the screen and feet at the bottom. In color television, another PLL makes sure that green remains green and red remains red (even if the politicians claim that the reverse is true).

A PLL is a circuit which causes a particular system to track with another one. More precisely, a PLL is a circuit synchronizing an output signal (generated by an oscillator) with a reference or input signal in frequency as well as in phase. In the synchronized—often called *locked*—state the phase error between the oscillator's output signal and the reference signal is zero, or very small.

If a phase error builds up, a control mechanism acts on the oscillator in such a way that the phase error is again reduced to a minimum. In such a control system the phase of the output signal is actually locked to the phase of the reference signal. This is why it is referred to as a *phase-locked loop.*

It is quite simple to deduce the operating principle of a PLL. Its block diagram is shown in Fig. 1-1$a$. The PLL consists of three basic functional blocks:

1. A voltage-controlled oscillator (VCO)

2. A phase detector (PD)

3. A loop filter (LF)

In some PLL circuits a current-controlled oscillator (CCO) is used instead of the VCO. In this case the output signal of the phase detector is a controlled current source rather than a voltage source. However, the operating principle remains the same.

The signals of interest within the PLL circuit are defined as follows:

· The reference (or input) signal $u_1(t)$

· The angular frequency $\omega_1$ of the reference signal

· The output signal $u_2(t)$ of the VCO

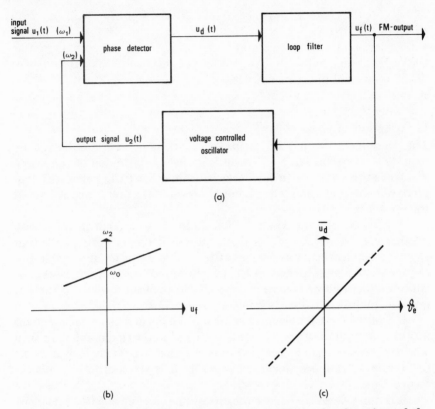

**Fig. 1-1** (*a*) **Block diagram of the PLL.** (*b*) **Transfer function of the VCO.** ($u_f$ = control voltage; $\omega_2$ = angular frequency of the output signal.) (*c*) **Transfer function of the PD.** ($\overline{u_d}$ = average value of the phase-detector output signal; $\theta_e$ = phase error.)

- The angular frequency $\omega_2$ of the output signal
- The output signal $u_d(t)$ of the phase detector
- The output signal $u_f(t)$ of the loop filter
- The phase error $\theta_e$, defined as the phase difference between signals $u_1(t)$ and $u_2(t)$

Let us now look at the operation of the three functional blocks in Fig. 1-1*a*. The VCO oscillates at an angular frequency $\omega_2$, which is determined by the output signal $u_f$ of the loop filter. The angular frequency $\omega_2$ is given by

$$\omega_2(t) = \omega_0 + K_0 u_f(t) \tag{1-1}$$

where $\omega_0$ is the center (angular) frequency of the VCO and $K_0$ is the VCO gain in $s^{-1} V^{-1}$.

Equation (1-1) is plotted graphically in Fig. 1-1$b$. In many textbooks the physical unit rad $s^{-1} V^{-1}$ is used for the VCO gain, since the unit rad $s^{-1}$ is often used for angular frequencies. We shall drop the unit radians in this text. (Note, however, that any phase variables used in this book will have to be measured in *radians* and not in *degrees!*) Therefore, in the equations a phase shift of 180° must always be specified as a value of $\pi$.

The PD—also referred to as *phase comparator*—compares the phase of the output signal with the phase of the reference signal and develops an output signal $u_d(t)$ which is approximately proportional to the phase error $\theta_e$, at least within a limited range of the latter

$$u_d(t) = K_d\theta_e \tag{1-2}$$

Here $K_d$ represents the gain of the PD. The physical unit of $K_d$ is volts. Some textbooks use the unit V rad$^{-1}$ for the reasons discussed above. Figure 1-1$c$ is a graphical representation of Eq. (1-2).

The output signal $u_d(t)$ of the PD consists of a dc component and a superimposed ac component. The latter is undesired, hence it is canceled by the loop filter. In most cases a first-order, low-pass filter is used. (Refer to Chap. 2.)

Let us now see how the three building blocks work together. First we assume that the angular frequency of the input signal $u_1(t)$ is equal to the center frequency $\omega_0$. The VCO then operates at its center frequency $\omega_0$. As we see, the phase error $\theta_e$ is zero. If $\theta_e$ is zero, the output signal $u_d$ of the PD must also be zero. Consequently the output signal of the loop filter $u_f$ will also be zero. This is the condition that permits the VCO to operate at its center frequency.

If the phase error $\theta_e$ were not zero initially, the PD would develop a nonzero output signal $u_d$. After some delay the loop filter would also produce a finite signal $u_f$. This would cause the VCO to change its operating frequency in such a way that the phase error finally vanishes.

Assume now that the frequency of the input signal is changed suddenly at time $t_0$ by the amount $\Delta\omega$. As shown in Fig. 1-2, the phase of the input signal then starts leading the phase of the output signal. A phase error is built up and increases with time. The PD develops a signal $u_d(t)$, which also increases with time. With a delay given by the loop filter, $u_f(t)$ will also rise. This causes the VCO to increase its frequency. The phase error becomes smaller now, and after some settling time the VCO will oscillate at a frequency that is exactly the frequency of the input signal. Depending on the type of loop filter used, the final phase error will have been reduced to zero or to a finite value.

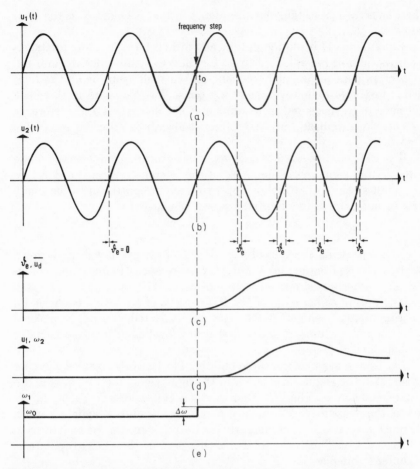

**Fig. 1-2  Transient response of a PLL onto a step variation of the reference frequency.** (*a*) **Reference signal** $u_1$ (*t*). (*b*) **Output signal** $u_2(t)$ **of the VCO.** (*c*) **Signals** $\overline{u_d}(t)$ **and** $\theta_e(t)$ **as a function of time.** (*d*) **Angular frequency** $\omega_2$ **of the VCO as a function of time.** (*e*) **Angular frequency** $\omega_1$ **of the reference signal** $u_1(t)$.

The VCO now operates at a frequency which is greater than its center frequency $\omega_0$ by an amount $\Delta\omega$. This will force the signal $u_f(t)$ to settle at a final value of $u_f = \Delta\omega/K_0$. If the center frequency of the input signal is frequency-modulated by an arbitrary low-frequency signal, then the output signal of the loop filter is the *demodulated signal*. The PLL can consequently be used as an (FM) detector. As we shall see later, it can be further applied as an AM or PM detector.

One of the most intriguing capabilities of the PLL is its ability to suppress noise superimposed on its input signal. Let us suppose that the input signal of the PLL is buried in noise. The PD tries to measure the phase error between input and output signals. The noise at the input causes the zero crossings of the input signal $u_1(t)$ to be advanced or delayed in a stochastic manner. This causes the PD output signal $u_d(t)$ to jitter around an average value. If the corner frequency of the loop filter is low enough, almost no noise will be noticeable in the signal $u_f(t)$, and the VCO will operate in such a way that the phase of the signal $u_2(t)$ is equal to the average phase of the input signal $u_1(t)$. Therefore we can state that the PLL is able to detect a signal that is buried in noise. These simplified considerations have shown that the PLL is nothing but a servo system which controls the phase of the output signal $u_2(t)$.

As shown in Fig. 1-2 the PLL was always able to track the phase of the output signal to the phase of the reference signal; this system was locked at all times. This is not necessarily the case, however, because a larger frequency step applied to the input signal could cause the system to "unlock." The control mechanism inherent in the PLL will then try to become locked again, but will the system indeed lock again? We shall deal with this problem in the following chapters. Basically two kinds of problems have to be considered:

- The PLL is initially locked. Under what conditions will the PLL remain locked?

- The PLL is initially unlocked. Under what conditions will the PLL become locked?

If we try to answer these questions, we will notice that different PLLs behave quite differently in this regard. We find that there are some fundamentally different types of PLLs. Therefore we will first identify these various types.

# 2 CLASSIFICATION OF PLL TYPES

Most PLL systems now in use have similar types of VCOs or CCOs and utilize first-order loop filters. They can differ considerably, however, in the type of phase-detector circuit used. The properties of the phase-detector circuit have a strong influence on the dynamic performance of the PLL system. Hence we will have to deal first with the characteristics of various phase-detector circuits.

Table 2-1 gives an overview of the most frequently used PDs; there are linear and digital types. The linear types are built from circuits which have previously been applied in the field of analog computation. The digital types, however, are based on logic circuits such as the EXCLUSIVE-OR gate. Moreover, digital PDs operate on binary signals exclusively, which means that both the reference and the output signals should be square waves.

The phase-detector circuit labeled *type 1* in Table 2-1 is simply an analog multiplier, also called a *four-quadrant multiplier.*

Assume for the moment that both reference and output signals are sine-wave signals and have the same frequency $\omega_1$,*

$$u_1(t) = \hat{U}_{10} \sin (\omega_1 t + \theta_1) \tag{2-1}$$

$$u_2(t) = \hat{U}_{20} \cos (\omega_1 t + \theta_2)$$

The phase-detector output signal $u_d(t)$ is by definition the product of these two signals, expressed as

$$u_d(t) = ku_1u_2 = \frac{k\hat{U}_{10}\hat{U}_{20}}{2} [\sin (\theta_1 - \theta_2) + \sin (2\omega_1 + \theta_1 + \theta_2)] \tag{2-2}$$

where $\hat{U}_{10}$ and $\hat{U}_{20}$ are the amplitudes and $\theta_1$ and $\theta_2$ the phases of $u_1$ and $u_2$, respectively, and $k$ is a gain constant.

Equation (2-2) reveals that $u_d(t)$ is a superposition of a dc and an ac component. The ac component is almost completely filtered out by the loop filter. Therefore we will henceforth consider the dc or average component of $u_d$ only, which is given by

$$\overline{u_d} = K_d \sin \theta_e \tag{2-3}$$

---

*The choice of sine and cosine functions is arbitrary; for more details refer to Sec. 3-1.

# Table 2-1 The properties of various types of PDs

| 1 PD Type | 2 Signals | 3 Schematic diagram | 4 phase error $\overline{u_d}$ / Output signals $\overline{u_d}$ as a function of frequency error $\omega_{in} \cdot \omega_{in2}$ | 5 PD sensitive on | 6 Operating mode | 7 Can be cascaded with low pass filter Type ..... |
|---|---|---|---|---|---|---|
| **1** linear | $u_2$ can also be a square wave | 4 Quadrant-Multiplier | | Phase | linear | all |
| **1** in saturation | $u_1$ Sine or square wave $u_2$ | EXCLUSIVE OR = | given by phase error alone | Phase | quasi digital | all |
| **2** | $u_1$, $u_2$ | EXCLUSIVE OR = | $\overline{u_d}$ = Duty cycle of the signal Q | Phase | digital | |
| **3** | $u_1$, $u_2$ | Edge triggered JK-Master-Slave-FF | undefined | Phase and frequency | digital | preferably cascaded with low pass filter type 3 having a pole at $\omega = 8$ (integrator) |
| **4** | Case 1: $U_1$ leading UP DOWN Case 2: $U_2$ leading UP DOWN | | $\overline{u_d}$ = weighted average of the outputs UP and DOWN UP: weight +1 DOWN: weight −1 | Phase and frequency | digital | |

$\overline{u_d}$ = weighted average of the outputs UP and DOWN
UP: weight +1   DOWN: weight −1

where

$$K_d = \frac{k\hat{U}_{10}\hat{U}_{20}}{2}$$

is the phase-detector gain and

$$\theta_e = \theta_1 - \theta_2$$

is the phase error.

Equation (2-3) is plotted graphically in col. 4 of Table 2-1. For small phase errors $\sin \theta_e$ is approximately equal to $\theta_e$, and $\overline{u_d}$ is approximately equal to $K_d\theta_e$. This is the phase-detector performance already postulated in Eq. (1-2). For small phase errors a PLL system utilizing a type 1 PD will operate as described in Chap. 1.

It should be mentioned that the PD gain $K_d$ as defined by Eq.(2-3) is dependent on the amplitude of the signals $u_1$ and $u_2$. This is clearly a disadvantage, since the dynamic performance of the PLL becomes dependent upon the amplitude of the reference signal. As long as the analog multiplier used in the type 1 PD operates in its linear region, the PD gain $K_d$ increases linearly with amplitude $\hat{U}_{10}$. For even larger reference signals the multiplier will become saturated, and $K_d$ will approach a limiting value, as shown in Fig. 2-1.

What about the performance of the type 1 PD if the system has not yet become locked? Then $\omega_1 \neq \omega_2$, and according to Eq. (2-3), the signal $u_d(t)$ will consist of two terms having angular frequencies of $\omega_1 - \omega_2$ and $\omega_1 + \omega_2$, respectively. There is no dc component now, which, at first glance, could lead to the (erroneous) conclusion that this system can never become locked. It is, however, possible if certain conditions are met. We shall discuss this phenomenon in detail in Chap. 3.

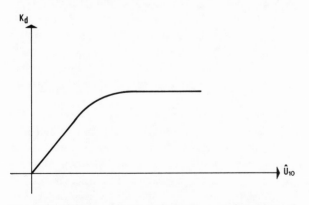

**Fig. 2-1 Phase-detector gain $K_d$ as a function of the amplitude $\hat{U}_{10}$ of the reference signal.**

9

Column 4 of Table 2-1 shows the average phase-detector output signal $\overline{u_d}$ both as a function of the phase error $\theta_e$ and as a function of the frequency offset $\omega_1 - \omega_2$. This second function is trivial, however, in the case of the type 1 PD. For this detector, the output signal is defined analytically for zero frequency offset only ($\omega_1 = \omega_2$). Note that this is different in the cases of the type 3 and type 4 PDs.

As we shall see later, the latter two types will become locked more easily when the initial frequency offset between output and reference signals is large. It would be incorrect, however, to state that the performance of the type 1 PD is inadequate, for it offers remarkable noise suppression. A noise signal superimposed on the reference signal can be thought of as a broad-band signal consisting of an infinite number of different frequency components. Because these frequencies are not correlated to the output signal of the VCO, the average output signal of the type 1 PD generated by these noise frequencies is zero. Hence a PLL system utilizing the type 1 PD is able to lock on signals which are heavily buried in noise. This explains the widespread use of PLLs in communication receivers.

In Eq. (2-2) we assumed that both the reference and the output signals are sine waves. However, in most practical PLLs the output signal of the VCO is a square wave. A symmetrical square-wave signal results in a frequency spectrum having only odd harmonics, that is, the angular frequencies are $3\omega_2$, $5\omega_2$, and so on. Only the fundamental frequency $\omega_2$ can contribute to a dc component of the signal $u_d(t)$; therefore the behavior of the type 1 PD is not dependent on the waveform of the VCO output signal.

If the reference signal of the PLL is made very large, the PD will operate in the saturated region. Its output signal $u_d(t)$ then indicates only two voltage levels, $U_+$ or $U_-$, which corresponds to the positive and negative saturation levels. In this case the average signal $\overline{u_d}$ depends only on the phase of the zero crossings of $u_1(t)$ with respect to $u_2(t)$. Moreover $\overline{u_d}$ is approximately proportional to the phase error $\theta_e$ in the overdriven state. This is plotted graphically in Table 2-1. It shows that the type 1 PD loses its ability to cancel noise signals when it is heavily overdriven. Consequently PDs used in communication receivers should always operate in the linear region.

Type 2, 3, and 4 PDs are *digital* circuits and require square waves for both reference and output signals. The simplest of these three devices is the EXCLUSIVE-OR gate used by the type 2 PD. The average signal $\overline{u_d}$ is given here by the duty-cycle ratio of the signal $Q$ at the output of the EXCLUSIVE-OR gate. If the signals $u_1$ and $u_2$ are exactly in phase, the output $Q$ is always zero and the average signal $\overline{u_d}$ is then nearly zero percent of the gate's supply voltage.

If $u_1$ and $u_2$ are out of phase by $\pi/2$ (90°), the $Q$ signal is a symmetrical square wave having twice the operating frequency of the PLL. The average signal $\overline{u_d}$ is then 50 percent of the supply voltage. If $u_1$ and $u_2$ are exactly opposite in phase, $Q$ will always be a logic 1, and $\overline{u_d}$ will be about 100 percent of the supply voltage. To get a symmetrical output signal, an offset of 50 percent of the supply voltage is usually added to the $Q$ signal. If we plot $\overline{u_d}$ as a function of phase error (Table 2-1, col. 4), we obtain a triangular curve which is symmetrical around the $\theta_e$ axis. Thus the type 2 PD behaves very much like the type 1 PD in its overdriven state. Note that the phase error $\theta_e$ is defined as zero if the phase offset is $\pi/2$ between signals $u_1$ and $u_2$; the same convention was established for the type 1 PD.

The triangular waveform of $\overline{u_d}$ vs. $\theta_e$ is obtained only if both the reference and output signals are symmetrical square waves. If one or both signals become asymmetrical, the function $\overline{u_d}$ vs. $\theta_e$ is clipped, as is represented by the dashed curve in Table 2-1, col. 4. This results in a lower phase-detector gain and reduces the lock range of the PLL, which will be discussed in Chap. 4.

The type 3 PD is simply an edge-triggered $JK$ master-slave flip-flop. In the circuit chosen here (Table 2-1, col. 3) the $Q$ output of the flip-flop is set HIGH by the falling edge of the signal $u_1$, and it is reset LOW by the falling edge of the signal $u_2$. The $\overline{Q}$ output is the logic complement of the $Q$ output. The average phase-detector signal $\overline{u_d}$ is defined here by the weighted duty-cycle ratio of the signals $Q$ and $\overline{Q}$: If $Q$ is HIGH, this corresponds to a weight of $+1$; if $\overline{Q}$ is HIGH, however, this corresponds to a weight of $-1$. The average output signal $\overline{u_d}$ is normally obtained by an additional averaging device, such as an $RC$ low-pass filter. In most cases this additional device is used simultaneously to implement the loop filter. These two functions are often implemented by a circuit called a *charge pump,* as shown in col. 7 of Table 2-1. This is essentially an integrator whose output signal goes up during the time when $Q$ is HIGH and moves down when $\overline{Q}$ is HIGH. If we plot the average signal $\overline{u_d}$ as a function of phase error, we get a sawtooth-like waveform (Table 2-1, col. 4). The function is periodic with modulo $2\pi$. It should be mentioned that this function is not dependent on the duty cycle of the square waves $u_1(t)$ and $u_2(t)$. This was different in the case of the type 2 PD.

In the unlocked state the dynamic performance of the type 3 PD differs considerably from that of types 1 and 2. If the frequency offset $\omega_1 - \omega_2$ is large, the average value of the phase-detector output signal $\overline{u_d}$ is not zero; it is positive for $\omega_1 > \omega_2$ and negative for $\omega_1 < \omega_2$. This behavior is highly desirable, because the $\overline{u_d}$ signal will cause the frequency of the VCO to be pulled toward the frequency of the reference signal. The type 3 PD is therefore said to be *phase- and frequency-sensitive.* Unfortunately, this frequency-sensitive performance is apparent for large frequency offsets only; a detailed analysis of the type 3 phase detector will be found in Sec. 4-1-1.

An analysis of the type 4 PD shows that this circuit outperforms type

3 in frequency-sensitive behavior. The type 4 PD is frequency-sensitive over the full range of frequency offset $\omega_1 - \omega_2$. Hence it is often referred to as a *phase/frequency detector*. As shown in the circuit diagram in col. 3 of Table 2-1, the NAND gates $G_1$ and $G_2$ form a latch labeled UP, whereas the NAND gates $G_3$ and $G_4$ form a latch called DOWN. The output signals of these two latches are usually taken to control a charge pump, as depicted in col. 7 of Table 2-1. The UP signal corresponds to a weight of $+1$, the DOWN signal to a weight of $-1$, as in the case of the type 3 PD. The UP and DOWN signals are defined here as *active-low* signals. Table 2-1, col. 2, shows the waveforms of the UP and DOWN signals for different values of phase error. In the trivial case of zero phase error (not shown in the table) both UP and DOWN signals are permanently HIGH. If the VCO output signal $u_2(t)$ lags the reference signal $u_1(t)$ (shown in case 1 of Table 2-1), the UP output generates pulses with a duty-cycle ratio proportional to the phase error $\theta_e$. The DOWN signal is then permanently HIGH or inactive. In the opposite case (case 2 in Table 2-1) $u_2(t)$ leads $u_1(t)$, the DOWN output is then pulsed, and the UP signal remains inactive. If the PLL system is not yet locked and $\omega_1$ is larger than $\omega_2$, the UP latch is more often set to its active state than the DOWN latch. Consequently the average value of $u_d(t)$ becomes positive. The reverse is true if $\omega_1$ is smaller than $\omega_2$. Note that the type 4 PD is frequency-sensitive for any value of the frequency offset $\omega_1 - \omega_2$. A more detailed analysis of the type 4 PD will be found in Sec. 4-1-1.

The dynamic performance of the PLL is influenced not only by the type of PD chosen, but also—even though less markedly—by the type of loop filter used in a particular application. In most cases the loop filter will be given by a first-order, low-pass filter. Thus we can restrict our dis-

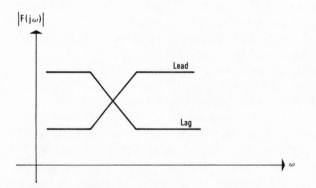

**Fig. 2-2  Bode diagrams of the realizable first-order loop filters according to Eq. (2-4).**

## Table 2-2 Possible first-order low-pass filters

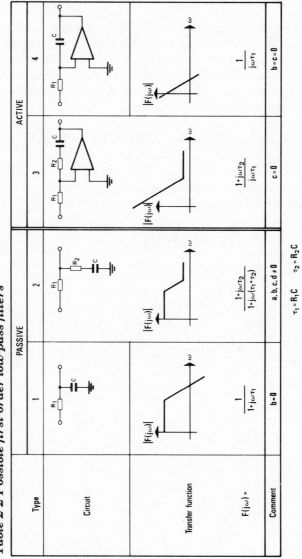

| Type | PASSIVE | | ACTIVE | |
|---|---|---|---|---|
| | 1 | 2 | 3 | 4 |
| Circuit | $R_1$, $C$ | $R_1$, $R_2$, $C$ | $R_1$, $R_2$, $C$ | $R_1$, $C$ |
| Transfer function $|F(j\omega)|$ | | | | |
| $F(j\omega) =$ | $\dfrac{1}{1+j\omega\tau_1}$ | $\dfrac{1+j\omega\tau_2}{1+j\omega(\tau_1+\tau_2)}$ | $\dfrac{1+j\omega\tau_2}{j\omega\tau_1}$ | $\dfrac{1}{j\omega\tau_1}$ |
| Comment | $b=0$ | $a,b,c,d \neq 0$ | $c=0$ | $b=c=0$ |

$$\tau_1 = R_1C \qquad \tau_2 = R_2C$$

13

cussion to this filter type for the moment. The most general form of transfer function for a first-order filter is given by

$$F(j\omega) = \frac{U_f(j\omega)}{U_d(j\omega)} = \frac{a + b(j\omega)}{c + d(j\omega)} \tag{2-4}$$

According to the location of the poles and zeroes, this function can represent either a low-pass or a high-pass filter (Fig. 2-2). Since the high-pass filter is useless for our purpose, we will concentrate on low-pass filters only. As shown in Table 2-2, there are four different practical realizations of a first-order, low-pass filter; these are designated types 1 to 4. These four types are characterized as follows:

Type 1: Passive $RC$ filter without a zero ($b = 0$).

Type 2: Passive $RC$ filter with a pole and a zero ($a, b, c, d \neq 0$).

Type 3: Active $RC$ filter having a pole at $\omega = 0$ ($c = 0$). This filter acts as an integrator at low frequencies.

Type 4: Active $RC$ filter without a zero ($b = c = 0$). This corresponds to an "ideal" integrator.

When designing a PLL system, we are free to combine any type of PD with any type of realizable loop filter. This does not simplify our theory, for different types of subsystems will lead to different performances of the overall PLL system. We should be careful when applying the techniques of one system to another one, as errors can easily result. This is the main reason why we deal with the theories of linear and digital PLLs separately.

# 3 Theory of the Linear PLL

It was shown in Chap. 1 that a PLL is simply a servo system controlling the phase $\theta_2$ of its output signal in such a way that the phase error between output phase $\theta_2$ and reference phase $\theta_1$ reduces to a minimum. We said that the PLL system was "in lock" (or "locked") when it was able to maintain phase tracking at all times.

When the power supply of a PLL is switched on for the first time, the VCO will barely oscillate at exactly the frequency of the reference signal. An acquisition process will therefore take place first. A locked state will be obtained if certain conditions are met; we shall discuss the lock-in process in detail in Sec. 3-1-2.

Once locked, the PLL will try to remain locked forever. This will be possible only if the variations of the phase or frequency of the reference signal are kept within given limits. We shall further discuss in Sec. 3-1-2 the requirements for keeping a PLL system *in lock, once it is locked.* It will be shown in Sec. 3-4 how the acquisition and tracking performance is influenced by noise signals at the PLL's reference input.

The term *linear PLL* means that this system is utilizing a linear type of PD, that is, the PD shown as *type 1* in Table 2-1. (From the view of control theory, however, the linear PLL is by no means a linear system!) An exact mathematical treatment of the PLL is very difficult. However, the network can be linearized for the discussion of tracking performance. The lock-in process on the other hand is always governed by nonlinear differential equations. Fortunately, we can find a quite simple mechanical analogy to the PLL and so greatly simplify the discussion of the acquisition process.

## 3-1 MATHEMATICAL FUNDAMENTALS

The following discussion considers the transient responses of the PLL system to various modulations of the reference signal. Because the PLL is a servo system which tries to track phase $\theta_2$ of the output signal with phase $\theta_1$ of the reference signal, the exciting function applied to the reference input will have to be expressed as a variation of the reference phase $\theta_1$ and not as a variation of the input voltage or current. The transient response of the PLL will then be obtained as a variation of output phase $\theta_2$. However, working with phase signals is unfamiliar to most electronic engineers. Let us illustrate what is meant by phase signals with the following simple examples.

Assume first that the reference signal of a PLL is a sine-wave signal of the form

$$u_1(t) = \hat{U}_{10} \sin [\omega_0 t + \theta_1(t)]$$

This signal is shown in Fig. 3-1*a*. Note that here the phase $\theta_1$ is defined as a function of time $t$. We assume furthermore that $\theta_1$ is 0 for $t < 0$ and has a value of $\Delta\Phi$ for $t \geq 0$. As is seen in Fig.3-1*a*, a phase step takes place at $t = 0$. The phase of the reference signal is therefore a *unit step*. So we can write

$$\theta_1(t) = \Delta\Phi \, u(t)$$

where $u(t)$ is the unit-step function. The signal considered in this example is a special case of phase modulation.

Let us also consider an example of frequency modulation (Fig. 3-1*b*). Assume the angular frequency of the reference signal is $\omega_0$ for $t < 0$. At $t = 0$ the angular frequency is abruptly changed by the increment $\Delta\omega$. For $t \geq 0$ the reference signal is consequently given by

$$u_1(t) = \hat{U}_{10} \sin (\omega_0 t + \Delta\omega t) = \hat{U}_{10} \sin (\omega_0 t + \theta_1)$$

In this case the phase $\theta_1$ can be written as

$$\theta_1(t) = \Delta\omega \, t$$

For FM reference signals the phase signal $\theta_1$ is therefore a ramp.

As a last example, consider a reference signal whose angular frequency is $\omega_0$ for $t < 0$ and increases linearly with time for $t \geq 0$ (Fig. 3-1*c*). For $t \geq 0$ its angular frequency is therefore

$$\omega_1(t) = \omega_0 + \dot{\Delta\omega} \, t$$

where $\dot{\Delta\omega}$ denotes the rate of change of angular frequency. Remember that the angular frequency of a signal is defined as the first derivative of its phase with respect to time:

$$\omega_1 = \frac{d\theta_1}{dt}$$

Hence the phase of a signal at time $t$ is the integral of its angular frequency over the time interval $0 \leq \tau \leq t$, where $\tau$ denotes elapsed time. The reference signal can be written as

$$u_1(t) = \hat{U}_{10} \sin \int_0^t (\omega_0 + \dot{\Delta\omega} \, \tau) \, d\tau$$

$$= \hat{U}_{10} \sin \left( \omega_0 t + \dot{\Delta\omega} \frac{t^2}{2} \right)$$

Consequently the corresponding phase signal $\theta_1(t)$ is given by

$$\theta_1(t) = \dot{\Delta\omega} \frac{t^2}{2}$$

In the following sections we shall calculate the output phase $\theta_2(t)$ for the most important cases of phase and frequency modulation of the reference signal. The output signal of the VCO is usually written in the form

$$u_2(t) = \hat{U}_{20} \cos [\omega_0 t + \theta_2(t)]$$

The choice of a cosine instead of a sine function is arbitrary. If a linear PLL is operating at its center frequency, there is a phase shift of $\pi/2$ (90°) between reference and output signals. If these two signals are defined as a sine and a cosine function, respectively, the phase error $\theta_e = \theta_1 - \theta_2$ then becomes exactly zero. Instead of calculating the output phase $\theta_2(t)$, we could derive an equation for the phase error $\theta_e(t)$ as an alternative. This calculation can be performed in either the time or the frequency domain.

### 3-1-1 Tracking Performance in the Frequency Domain

3-1-1-1 Linearized Model of the PLL    In classical control theory the dynamic performance of a system is generally discussed in the complex frequency domain by applying the Laplace transform. (Appendix F is an introduction to the Laplace transform.) The Laplace transform can be used for linear systems only. However, as shown in Chap. 2, the type 1 PD exhibits nonlinearity because the average output signal is a sine function of the phase error:

$$\overline{u_d} = K_d \sin \theta_e$$

If we postulate that the PLL system stays locked at all times and that the phase error $\theta_e$ remains relatively small, we can write $\sin \theta_e \approx \theta_e$. For practical purposes this approximation is valid for quite large phase errors, even for phase errors in the region of $\pi/3$ (corresponding to 60°). The system can then be considered linear. Let us keep in mind that the following considerations are valid only for the locked state.

Throughout this text we will use lowercase symbols for signals that are a function of time and uppercase letters for the corresponding Laplace transforms. The same rule will also apply to Greek letters.

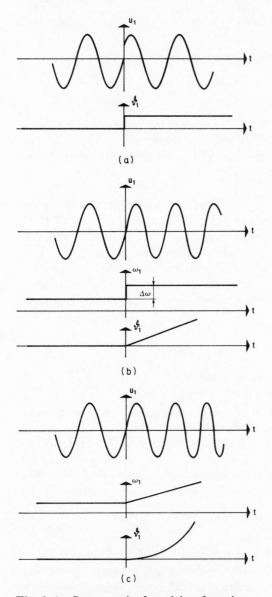

**Fig. 3-1 Some typical exciting functions as applied to the reference input of a PLL.** (*a*) **Phase error applied at** $t = 0$; $\theta_1 = \Delta\Phi\, u(t)$. (*b*) **Frequency step** $\Delta\omega$ **applied at** $t = 0$; $\theta_1 = \Delta\omega\, t$. (*c*) **Frequency ramp starting at** $t = 0$; $\theta_1 = \dot{\Delta\omega}\, t^2/2$.

Let us introduce the following variables into the complex frequency domain:

$$U_d(s) = \mathcal{L}\{u_d(t)\}$$

$$U_f(s) = \mathcal{L}\{u_f(t)\}$$

$$\Theta_1(s) = \mathcal{L}\{\theta_1(t)\}$$

$$\Theta_2(s) = \mathcal{L}\{\theta_2(t)\}$$

$$\Theta_e(s) = \mathcal{L}\{\theta_e(t)\}$$

where $s$ is the Laplace operator or complex (angular) frequency.

We can now draw a linearized block diagram of the PLL in the complex frequency domain (Fig. 3-2). The three functional blocks of the PLL have the following transfer functions:

· Phase detector: $\qquad \dfrac{U_d(s)}{\Theta_e(s)} = K_d \qquad\qquad$ (3-1a)

· Loop filter: $\qquad \dfrac{U_f(s)}{U_d(s)} = F(s) \qquad\qquad$ (3-1b)

· VCO: $\qquad \dfrac{\Theta_2(s)}{U_f(s)} = \dfrac{K_0}{s} \qquad\qquad$ (3-1c)

The transfer function of the PD follows directly from Eq. (1-2). The transfer function of the loop filter has already been defined by Eq. (2-4).

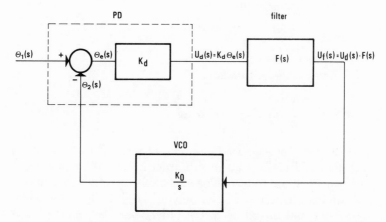

**Fig. 3-2 Linear model of the PLL. (Symbols defined in text.)**

19

The transfer function of the VCO can be derived from Eq. (1-1), which reads

$$\omega_2(t) = \omega_0 + K_0 u_f(t) \tag{1-1}$$

As stated in Sec. 3-1, the phase of a signal is equal to the time integral of angular frequency. Thus we can write

$$\omega_0 t + \theta_2(t) = \omega_0 t + K_0 \int_0^t u_f(t) \, dt$$

Applying the Laplace transform, we get

$$\Theta_2(s) = \frac{K_0 U_f(s)}{s}$$

which immediately leads to the transfer function of the VCO given in Eq. (3-1c).

From Eqs. (3-1) we can calculate either $\Theta_2(s)$ as a function of $\Theta_1(s)$ or $\Theta_e(s)$ as a function of $\Theta_1(s)$. For $\Theta_2(s)$ and $\Theta_e(s)$ we get

$$\Theta_2(s) = \Theta_1(s) \frac{K_0 K_d F(s)}{s + K_0 K_d F(s)} \tag{3-2a}$$

$$\Theta_e(s) = \Theta_1(s) \frac{s}{s + K_0 K_d F(s)} \tag{3-2b}$$

We now introduce two transfer functions which will be used very frequently throughout this text.

- Phase transfer function:

$$H(s) = \frac{\Theta_2(s)}{\Theta_1(s)} = \frac{K_0 K_d F(s)}{s + K_0 K_d F(s)} \tag{3-3}$$

- Error-transfer function:

$$H_e(s) = \frac{\Theta_e(s)}{\Theta_1(s)} = \frac{s}{s + K_0 K_d F(s)} \tag{3-4}$$

**3-1-1-2 First-, Second-, and Higher-Order Loops**  If the dynamic response of a system is described by an $n$th-order differential equation, the order of the system is said to be $n$. The denominator of the corresponding transfer function is then an $n$th-order polynomial in $s$. If we assume a first-order loop filter and introduce a first-order transfer function for $F(s)$ in Eq. (3-3) or Eq. (3-4), the order of the phase- and error-transfer functions becomes 2. We can state therefore that the order of the PLL system is equal to the order of the loop filter plus 1:

Order of PLL = order of loop filter + 1

The following discussion will consider mainly second-order PLLs, which are by far the type most commonly used. Such a system behaves very much like the popular *RLC* resonant circuit. Thus most important results may be taken from the theory of linear networks. In a few cases no loop filter is used at all within the PLL. The transfer function of the loop filter can then be considered to be $F(s) = 1$, which corresponds to a zero-order, low-pass filter. The resulting PLL then is first-order.

Higher-order loop filters have been implemented in a few applications. A second-order loop filter yields a third-order PLL. Because the overall phase shift of such a loop may exceed 180°, special means have to be provided to obtain stable operation. An example of a third-order PLL will be given in Sec. 7-9.

3-1-1-3 Introducing Normalized Variables   To obtain the phase-transfer function of the linear second-order PLL, we have to substitute the transfer function of the loop filter $F(s)$ into Eq. (3-3). In Table 2-2 we introduced four different transfer functions of the first-order loop filter. However, the type 1 loop filter is a special case of the type 2 filter with time constant $\tau_2 = 0$, and furthermore the type 4 loop filter is a special case of the type 3 filter with $\tau_2 = 0$. Consequently we have to perform the calculation for two cases only; the passive loop filter (types 1 and 2) and the active loop filter (types 3 and 4). For the phase-transfer function $H(s)$ we get:

· For a passive loop filter

$$H(s) = \frac{K_0 K_d (s\tau_2 + 1)/\tau_1}{s^2 + s\left(K_0 K_d \dfrac{\tau_2}{\tau_1}\right) + \dfrac{K_0 K_d}{\tau_1}} \tag{3-5a}$$

· For an active loop filter

$$H(s) = \frac{K_0 K_d (s\tau_2 + 1)/(\tau_1 + \tau_2)}{s^2 + s\left(\dfrac{1 + K_0 K_d \tau_2}{\tau_1 + \tau_2}\right) + \dfrac{K_0 K_d}{\tau_1 + \tau_2}} \tag{3-5b}$$

In circuit and control theory it is common practice to write the denominator of expressions such as Eqs.(3-5) in the so-called *normalized form*,

Denominator $= s^2 + 2\zeta\omega_n s + \omega_n^2$

where $\omega_n$ is the *natural frequency* and $\zeta$ is the *damping factor*. The denominator in Eqs. (3-5) will take this form if the following substitutions are made.

· For the active loop filter:

$$\omega_n = \left(\frac{K_0 K_d}{\tau_1}\right)^{1/2}$$

$$\zeta = \frac{\tau_2}{2}\left(\frac{K_0 K_d}{\tau_1}\right)^{1/2}$$

(3-6a)

· For the passive loop filter:

$$\omega_n = \left(\frac{K_0 K_d}{\tau_1 + \tau_2}\right)^{1/2}$$

$$\zeta = \frac{1}{2}\left(\frac{K_0 K_d}{\tau_1 + \tau_2}\right)^{1/2}\left(\tau_2 + \frac{1}{K_0 K_d}\right)$$

(3-6b)

(The natural frequency $\omega_n$ must never be confused with the center frequency $\omega_0$ of the PLL.)

Applying the substitution shown in Eqs. (3-6), we get the following phase-transfer functions.

· For an active loop filter:

$$H(s) = \frac{2\zeta\omega_n s + \omega_n^2}{s^2 + 2\zeta\omega_n s + \omega_n^2}$$

(3-7a)

· For a passive loop filter:

$$H(s) = \frac{s\omega_n\left(2\zeta - \dfrac{\omega_n}{K_0 K_d}\right) + \omega_n^2}{s^2 + 2\zeta\omega_n s + \omega_n^2}$$

(3-7b)

Aside from the parameters $\omega_n$ and $\zeta$, only one further parameter, $K_0 K_d$, is left in Eqs. (3-7). The term $K_0 K_d$ is called the *loop gain* and has the dimension of angular frequency $(\text{s}^{-1})$. If the condition

$$K_0 K_d \gg \omega_n$$

is true, this PLL system is said to be a *high-gain loop*. If the reverse is true, the system is called a *low-gain loop*. Most practical PLLs are high-gain loops. If a PLL using a passive loop filter is a high-gain loop, its phase-transfer function can be approximated as

$$H(s) \approx \frac{2s\zeta\omega_n + \omega_n^2}{s^2 + 2\zeta\omega_n s + \omega_n^2} \tag{3-8}$$

This is the same expression as that obtained for the PLL utilyzing an active filter.

To investigate the transient response of a control system, it is customary to plot a Bode diagram of its transfer function. The Bode diagram of the phase-transfer function is obtained by putting $s = j\omega$ in Eqs. (3-7) and by plotting the magnitude (absolute value) $|H(j\omega)|$ as a function of angular frequency $\omega$ (Fig. 3-3). Both scales are usually logarithmic. The frequency scale is further normalized to the natural frequency $\omega_n$. Thus the graph is valid for every second-order PLL system. We can see from Fig. 3-3 that the second-order PLL is actually a low-pass filter for input phase signals $\theta_1(t)$ whose frequency spectrum is between zero and approximately the natural frequency $\omega_n$. This means that the second-order PLL is able to track for phase and frequency modulations of the reference sig-

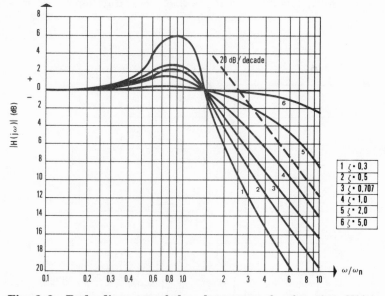

**Fig. 3-3** **Bode diagram of the phase-transfer function $H(j\omega)$.** *(Adapted from Gardner[1] with permission.)*

nal as long as the modulation frequencies remain within an angular frequency band roughly between zero and $\omega_n$.

The damping factor $\zeta$ has an important influence on the dynamic performance of the PLL. For $\zeta = 1$ the system is critically damped. If $\zeta$ is made smaller than unity, the transient response becomes oscillatory; the smaller the damping factor, the larger becomes the overshoot. In most practical systems, an optimally flat frequency-transfer function is the goal. The transfer function is optimally flat for $\zeta = 1/\sqrt{2} \approx 0.7$, which corresponds to a second-order Butterworth low-pass filter. If $\zeta$ is made considerably larger than unity, the transfer function flattens out, and the dynamic response becomes sluggish.

The dynamic performance of the PLL can also be considered by means of the error-transfer function $H_e(s)$. If we substitute the transfer function of a first-order low-pass filter into Eq. (3-4), and assume a high-gain loop, we get

$$H_e(s) \approx \frac{s^2}{s^2 + 2s\zeta\omega_n + \omega_n^2} \tag{3-9}$$

This expression is again identical for PLLs using an active loop filter and for high-gain PLLs having a passive loop filter.

A Bode plot of $H_e(s)$ is shown in Fig. 3-4. The value of 0.707 has been chosen for $\zeta$. The diagram shows that for modulation frequencies smaller

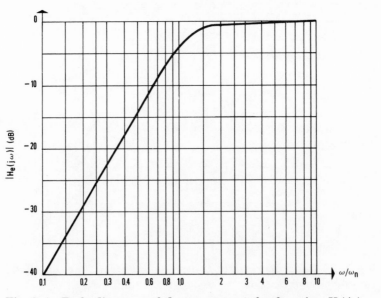

**Fig. 3-4  Bode diagram of the error-transfer function $H_e(j\omega)$.**

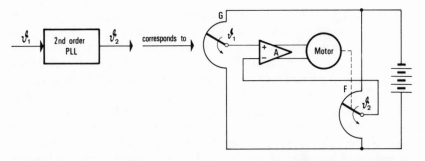

**Fig. 3-5 Simple electromechanical analogy of the linearized second-order PLL. In this servo system, the angles $\theta_1$ and $\theta_2$ correspond to the phases $\theta_1$ and $\theta_2$, respectively, of the PLL system.**

than the natural frequency $\omega_n$ the phase error remains relatively small. For larger frequencies, however, the phase error $\theta_e$ becomes as large as the reference phase $\theta_1$, which means that the PLL is no longer able to maintain phase tracking.

Knowing that a second-order PLL in the locked state behaves very much like a servo or follow-up control system, we can plot a simple model for the locked PLL (Fig. 3-5). The model consists of a reference potentiometer $G$, a servo amplifier, and a follow-up potentiometer $F$ whose shaft is driven by an electric motor. In this model the reference phase $\theta_1$ is represented by the shaft position of the reference potentiometer $G$. The phase of the output signal of the VCO $\theta_2(t)$ is represented by the shaft position of the follow-up potentiometer. If the shaft position of the reference potentiometer is varied slowly, the servo system will be able to maintain tracking of the follow-up potentiometer. If $\theta_1(t)$ is changed too abruptly, the servo system will lose tracking and large phase errors $\theta_e$ will result.

So far we have seen that a linear model is best suited to explain the tracking performance of the PLL *if it is assumed that the PLL is initially locked.* If the PLL is initially unlocked, however, the phase error $\theta_e$ can take on arbitrarily large values, and the linear model is no longer valid. When we try to calculate the *acquisition process* of the PLL itself, we must use a model which also accounts for the nonlinear effect of the phase detector. This is done in the following section.

### 3-1-2 Calculation of Acquisition in the Time Domain

When trying to find a mathematical solution for the lock-in process, we must assume that the PLL is initially unlocked. The linear model is then invalid, so the Laplace transform cannot be used. The mathematical treatment must therefore be performed in the time domain. It starts with writing the nonlinear differential equation of the PLL in the unlocked state and continues with looking for an explicit solution. Unfortunately there is no exact mathematical solution for the acquisition process. However, it is not difficult to solve the nonlinear differential equation by numerical integration on a digital computer. We shall discuss some computer solutions for the lock-in process in Sec. 3-2. Instead of looking for numerical solutions, we can find a mechanical analogy of the PLL which is described by the same type of nonlinear differential equation. Because the mechanical analogy will turn out to be a very simple one, it will make it much easier to explain the nonlinear phenomena related to the lock-in process. Moreover, we shall find that many of the key parameters describing the lock-in process can be defined by means of the mechanical analogy. Although the mechanical analogy does not deliver numerical data for key parameters such as the lock-in range, the pull-out range, and so on, it gives some reasonable estimates. The mechanical analogy shows very clearly what kind of impact on the PLL is generated by a phase step, a frequency step, or a linear frequency sweep applied to the reference signal.

When analyzing the lock-in process, we should find a differential equation for the phase error $\theta_e(t)$. Its derivation is straightforward since the performance of the individual functional blocks of a PLL in the time domain is known.

From Eq. (2-3), the equation for the phase detector is

$$u_d(t) = K_d \sin \theta_e \tag{3-10}$$

For the following discussion we will choose a first-order type 2 loop filter as shown in Table 2-2. Its transfer function has been shown to be

$$F(j\omega) = \frac{1 + j\omega\tau_2}{1 + j\omega(\tau_1 + \tau_2)} = \frac{U_f(j\omega)}{U_d(j\omega)}$$

If this expression is transformed back into the time domain, a first-order differential equation for the output signal $u_f(t)$ of the loop filter is obtained:

$$u_f(t) + \dot{u}_f(t)(\tau_1 + \tau_2) = u_d(t) + \dot{u}_d(t)\tau_2 \tag{3-11}$$

The transfer function of the VCO has been shown in Eq. (3-1c) to be

$$\frac{\Theta_2(s)}{U_f(s)} = \frac{K_0}{s}$$

When we transform back into the time domain, the output phase $\theta_2(t)$ becomes

$$\theta_2(t) = K_0 \int_0^t u_f(t) \, dt \tag{3-12}$$

Equations (3-10), (3-11), and (3-12) are a system of differential equations for the three variables $\theta_e$, $u_d$, and $u_f$. After eliminating the variables $u_d$ and $u_f$ we obtain the nonlinear differential equation for the phase error $\theta_e$,

$$\ddot{\theta}_e + \dot{\theta}_e \frac{1 + K_0 K_d \tau_2 \cos \theta_e}{\tau_1 + \tau_2} + \frac{K_0 K_d}{\tau_1 + \tau_2} \sin \theta_e = \ddot{\theta}_1 + \dot{\theta}_1 \frac{1}{\tau_1 + \tau_2} \qquad (3\text{-}13)$$

The substitutions of Eqs. (3-6) are now introduced for $\tau_1$ and $\tau_2$. Practical considerations show that the inequality

$$\frac{1}{\tau_2} \ll K_0 K_d$$

is valid for most practical cases. This leads to the simplified differential equation

$$\ddot{\theta}_e + 2\zeta \omega_n \dot{\theta}_e \cos \theta_e + \omega_n^2 \sin \theta_e = \ddot{\theta}_1 + \dot{\theta}_1 \frac{\omega_n^2}{K_0 K_d} \qquad (3\text{-}14)$$

The nonlinearities in this equation stem from the trigonometric terms $\sin \theta_e$ and $\cos \theta_e$.

As already stated, there is no exact solution for this problem. We find, however, that Eq. (3-14) is almost identical to the differential equation of a somewhat special mathematical pendulum, as shown in Fig. 3-6. A beam having a mass $M$ is rigidly fixed to the shaft of a cylinder which can rotate freely around its axis. A thin rope is attached at point $P$ to the surface of the cylinder and is then wound several times around the latter. The other end of the rope hangs down freely and is attached to a weighing platform. If there is no weight on the platform, the pendulum is assumed to be in a vertical position with $\Phi_e = 0$. If some weight $G$ is placed on the platform, the pendulum will be deflected from its quiescent position and eventually settle at a final deflection angle $\Phi_e$. The dynamic response of the pendulum can be calculated by Newton's third law,

$$T\ddot{\Phi}_e = \sum_i J_i \qquad (3\text{-}15)$$

where $T$ is the moment of inertia of the pendulum plus cylinder, $\Phi_e$ is the angle of deflection, and $J_i$ is a driving torque. Three different torques can be identified in the mechanical system of Fig. 3-6:

1. The torque $J_E$ generated by gravitation of the mass $M$; $J_E = -Ma\boldsymbol{g} \sin \Phi_e$, where $a$ is the length of the beam and $\boldsymbol{g}$ is acceleration due to gravity.

2. A friction torque $J_R$, which is assumed to be proportional to the

27

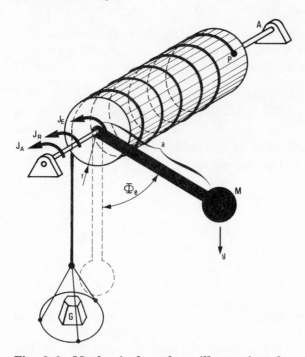

**Fig. 3-6 Mechanical analogy illustrating the linear and the nonlinear performance of the PLL. (Symbols defined in text.)**

angular velocity $\dot{\Phi}_e$ (viscous friction); $J_R = -\rho\dot{\Phi}_e$, where $\rho$ is the coefficient of friction.

3. The torque $J_A$ generated by the gravity of the weight $G$; $J_A = rG$, where $r$ is the radius of the cylinder.

Introducing these individual torques into Eq. (3-15) yields

$$\ddot{\Phi}_e + \frac{\rho}{T}\dot{\Phi}_e + \frac{Mag}{T}\sin\Phi_e = \frac{r}{T}G(t) \tag{3-16}$$

As in the case of the PLL, we can write this equation in a normalized form. If we introduce the substitutions

$$\omega_n' = \left(\frac{Mag}{T}\right)^{1/2}$$

$$\tag{3-17}$$

$$\zeta' = \frac{\rho}{2\sqrt{Mag\,T}}$$

Equation (3-16) is converted into

$$\ddot{\Phi}_e + 2\zeta'\omega_n'\dot{\Phi}_e + \omega_n'^2 \sin \theta_e = \frac{r}{T} G(t) \sim G(t) \qquad (3\text{-}18)$$

This nonlinear differential equation for the deflection angle $\Phi_e$ looks very much like the nonlinear differential equation of the PLL according to Eq. (3-13). There is a slight difference in the second term, however. In the case of the PLL the second term contains the factor $\cos \theta_e$, whereas for the pendulum the coefficient of the second term $(\dot{\Phi}_e)$ is the constant $2\zeta'\omega_n'$. Strictly speaking, the pendulum of Fig. 3-6 would only be an accurate analogy of the PLL if the friction varied with the cosine of the deflection angle. This would be true if the damping factor $\zeta'$ were not a constant but would vary with $\cos \Phi_e$. As a consequence, the momentary friction torque would be positive for

$$-\frac{\pi}{2} < \Phi_e < +\frac{\pi}{2}$$

that is, when the position of the pendulum is in the lower half of a circle around the cylinder shaft. On the other hand, the momentary friction torque would be negative for

$$\frac{\pi}{2} < \Phi_e < \frac{3\pi}{2}$$

that is, when the position of the pendulum is in the upper half of this circle. A negative friction is hard to imagine, of course, but let us assume for the moment that $\zeta' \sim \cos \Phi_e$ is valid. Imagine further that the weight $G$ is large enough to make the pendulum tip over and continue to rotate forever around its axis (provided the rope is long enough). Because of the nonconstant torque generated by the mass $M$ of the pendulum this oscillation will be nonharmonic. During the time when the pendulum swings through the lower half of the circle $(-\pi/2 \le \Phi_e \le \pi/2)$, its average angular velocity is greater than its velocity during the time when it swings through the upper half $(\pi/2 \le \Phi_e \le 3\pi/2)$. The *positive* friction torque averaged over the lower semi-circle is therefore greater in magnitude than the *negative* friction torque averaged over the upper semi-circle. This means that the friction torque averaged over one full revolution stays positive; hence it is acceptable to state that the coefficient of friction $\rho$ varies with the cosine of $\Phi_e$. The mathematical pendulum is therefore a reasonable approximation for the PLL.

Comparing the PLL with this mathematical pendulum, we find the following analogies:

1. The phase error $\theta_e$ of the PLL corresponds to the angle of deflection $\Phi_e$ of the pendulum.

2. The natural frequency $\omega_n$ of the PLL corresponds to the natural (or resonant) frequency $\omega'_n$ of the pendulum.

3. The damping factor $\zeta$ of the PLL corresponds to the damping factor $\zeta'$ of the pendulum, which results from viscous friction.

4. The weight $G$ on the platform corresponds to a reference phase disturbance according to the relation

$$\ddot{\theta}_1 + \dot{\theta}_1 \frac{\omega_n^2}{K_0 K_d} \sim G(t) \tag{3-19}$$

Let us now see what is the physical meaning of the term $\ddot{\theta}_1 + \dot{\theta}_1(\omega_n^2/K_0 K_d)$ in Eq. (3-19). We assume first that the frequency of the reference signal has an arbitrary value:

$$\omega_1 = \omega_0 + \Delta\omega(t)$$

where $\Delta\omega(t)$ can be considered the frequency offset of the reference signal. The reference signal $u_1(t)$ can therefore be written in the form

$$u_1(t) = \hat{U}_{10} \sin \left\{ \int_0^t [\omega_0 + \Delta\omega(t)] \, dt \right\}$$

$$= \hat{U}_{10} \sin [\omega_0 t + \theta_1(t)]$$

Consequently the phase $\theta_1(t)$ is given by

$$\theta_1(t) = \int_0^t \Delta\omega(t) \, dt$$

From this it becomes immediately evident that the first derivative $\dot{\theta}_1(t)$ represents the momentary frequency offset:

$$\dot{\theta}_1(t) = \Delta\omega(t)$$

whereas the second derivative $\ddot{\theta}_1(t)$ signifies the rate of change of the frequency offset:

$$\ddot{\theta}_1(t) = \frac{d}{dt} \Delta\omega(t) = \dot{\Delta\omega}$$

The weight $G$ placed on the platform is thus equivalent to a weighted sum of the frequency offset $\Delta\omega(t)$ and its rate of change $\dot{\Delta\omega}$:

$$G(t) \sim \dot{\Delta\omega} + \frac{\omega_n^2}{K_0 K_d} \Delta\omega \tag{3-20}$$

This simple correspondence paves the way toward understanding the quite complex dynamic performance of a PLL in the locked and unlocked states. To see what happens to a PLL when phase and/or frequency steps of arbitrary size are applied to its reference input, we have to place the corresponding weight $G(t)$ given by Eq. (3-20) on the platform and observe the response of the pendulum. The notation $G(t)$ should emphasize that $G$ must not necessarily be a constant, but can also be a function of time, as would be the case when an impulse is applied.

Let us first consider the trivial case of no weight on the platform. The pendulum is then at rest in a vertical position, $\Phi_e = 0$. This corresponds to the PLL operating at its center frequency $\omega_0$ with zero frequency offset ($\Delta\omega = 0$) and zero phase error ($\theta_e = 0$).

What happens if the frequency of the reference signal is changed *slowly?* The rate of change of the reference frequency is assumed to be so low that the derivative term $\dot{\Delta\omega}$ in Eq. (3-20) is negligible. A slow variation of the reference frequency corresponds to a slow increase of weight $G$, achieved by very carefully pouring a fine powder onto the platform. The analogy is given in this case by

$$G \sim \frac{\omega_n^2}{K_0 K_d} \Delta\omega$$

The pendulum now starts to deflect, indicating that a finite phase error is established within the PLL. For small offsets of the reference frequency the phase error $\theta_e$ will be proportional to $\Delta\omega$. If the frequency offset reaches a critical value, called the *hold range,* the deflection of the pendulum is just 90°. This is the *static limit of stability.*

With the slightest disturbance the pendulum would now tip over and rotate around its axis forever. This corresponds to the case where the PLL is no longer able to maintain phase tracking and consequently unlocks. One full revolution of the pendulum equals a phase error of $2\pi$. Because the pendulum is now rotating permanently, the phase error increases toward infinity.

Another interesting case is given by a step change of the reference frequency at the input of the PLL. When a frequency step of the size $\Delta\omega$ is applied at $t = 0$, the angular frequency of the reference signal is

$$\omega_1(t) = \omega_0 + \Delta\omega \, u(t)$$

where $u(t)$ is the unit-step function. The first derivative $\dot{\Delta\omega}$ therefore shows a delta function at $t = 0$; this is written

$$\dot{\Delta\omega}(t) = \Delta\omega \, \delta(t)$$

and is plotted in Fig. 3-7. What will now be the weight $G(t)$ required to simulate this condition? As also shown in Fig. 3-7, the weight function should be a superposition of a step function and a delta (impulse) function. In practice this can be simulated by dropping an appropriate weight

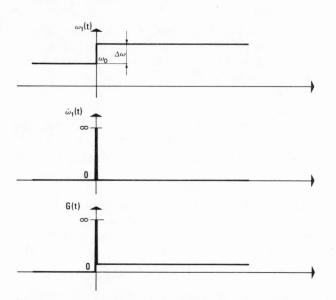

**Fig. 3-7** **Weight function** $G(t)$ **required to simulate a frequency step applied to the reference input of the PLL.** ($\omega_1$ = **angular frequency of the reference signal;** $\Delta\omega$ = **frequency step applied at** $t$ = **0.**)

from some height onto the platform. The impulse is generated when the weight hits the platform. To get a narrow and steep impulse, the stroke should be *elastic*. If this is done, the pendulum will show a transient response, mostly in the form of damped oscillation. If a relatively small weight is *dropped* onto the platform, the final deflection of the pendulum will be the same as if the weight had been placed *smoothly* onto the platform. If the pendulum is not heavily overdamped, however, its *peak deflection* $\hat{\Phi}_e$ will be considerably greater than its final deflection. If we increase the weight to be dropped onto the platform, we will observe a situation where the peak deflection exceeds 90°, but not 180°, and the final deflection is less than 90°. We thus conclude that a linear PLL can operate stably when the phase error *momentarily* exceeds the value of 90°. If the weight dropped onto the platform is increased even further, the peak deflection will exceed 180°. The pendulum now tips over and performs a number of revolutions around its axis, but it will probably come to rest again after some time. The *weight* which *caused the system to unlock* (at least temporarily) is observed to be considerably *smaller* than the weight that represented the *hold range*. We therefore have to define another critical-frequency offset, i.e., the offset that causes the PLL to unlock when it is applied as a step. This frequency step is called the *pull-out range.* Keep in mind that the pull-out range of a PLL is markedly smaller than its hold range. The pull-out range may be considered the *dynamic limit of stability.* The PLL always stays locked as long as the frequency steps applied to the system do not exceed the pull-out range.

There is a third way to cause a PLL to unlock from an initially stable operating point. If the frequency of the reference signal is increased linearly with time, we have

$$\Delta\omega = \dot{\Delta\omega} \, t$$

where $\dot{\Delta\omega}$ is the rate of change of the angular frequency. In the mechanical analogy this corresponds to a weight $G$ that also builds up linearly with time. This can be realized by feeding a material onto the platform at an appropriate feed rate. It becomes evident that too rapid a feed rate acts on the pendulum like the impulse which was generated in the last example by dropping a weight onto the platform. As will be seen later (Sec. 3-3-2-3), the rate of change of the reference frequency must always be smaller than $\omega_n^2$ to keep the system locked:

$$\dot{\Delta\omega} < \omega_n^2$$

These examples have demonstrated that three conditions are necessary if a PLL system is to maintain phase tracking:

1. The angular frequency of the reference signal must be within the hold range.

2. The maximum frequency step applied to the reference input of a PLL must be smaller than the pull-out range.

3. The rate of change of the reference frequency $\dot{\Delta\omega}$ must be lower than $\omega_n^2$.

Whenever a PLL has lost tracking because one of these conditions has not been fulfilled, the question arises whether it will return to stable operation when all the necessary conditions are met again. The answer is clearly *no.* When the reference frequency exceeded the hold range, the pendulum in our analogy tipped over and started rotating around its axis. If the weight on the platform were reduced to a value slightly less than the critical limit that caused instability, the pendulum would nevertheless continue to rotate because there is enough kinetic energy stored in the mass to maintain the oscillation. If there were no friction at all, the pendulum would continue to rotate even if the weight $G$ were reduced to zero. Fortunately friction exists, so the pendulum will decelerate if the weight is decreased to an appropriate level. For a PLL, this means that buildup of the phase error will decelerate if the reference-frequency offset is decreased below another critical value, the *pull-in frequency*. If the slope of the average phase error becomes smaller, the frequency $\omega_2$ of the VCO more and more approaches the frequency of the reference signal, and the system will finally lock. The pull-in frequency $\Delta\omega_P$ is markedly smaller than the hold range, as can be expected from the mechanical analogy. The pull-in process is a relatively slow one, as will be demonstrated in Sec. 3-2-3.

In most practical applications it is desired that the locked state be obtained within a short time period. Suppose again that the weight put on the platform of the mechanical model is large enough to cause sustained oscillation of the pendulum. It is easily shown that the pendulum can be brought to rest *within one single revolution* if the weight is suddenly decreased below another critical value. This implies that a PLL can become locked within one single-beat note between reference frequency and output frequency, provided the frequency offset $\Delta\omega$ is reduced below a critical value called the *lock range*. This latter process is called the *lock-in process*. The lock-in process is much faster than the pull-in process, but the lock range is smaller than the pull-in range.

The mechanical model has shown that there are four key parameters specifying the frequency range in which the PLL can be operated. Figure

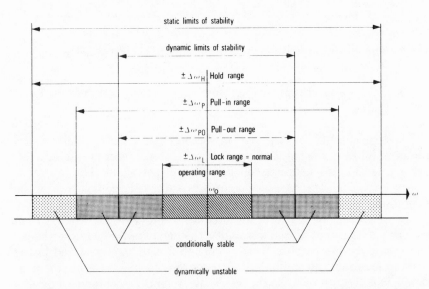

**Fig. 3-8  Scope of the static and dynamic limits of stability of a linear second-order PLL.**

3-8 is graphical representation of these parameters. The four key parameters can be summarized as follows:

1. The *hold range* $\Delta\omega_H$. This is the frequency range in which a PLL can statically maintain phase tracking. A PLL is conditionally stable only within this range.

2. The *pull-out range* $\Delta\omega_{PO}$. This is the dynamic limit for stable operation of a PLL. If tracking is lost within this range, a PLL normally will lock again, but this process can be slow if it is a *pull-in process*.

3. The *pull-in range* $\Delta\omega_P$. This is the range within which a PLL will always become locked, but the process can be rather slow.

4. The *lock range* $\Delta\omega_L$. This is the frequency range within which a PLL locks within one single-beat note between reference frequency and output frequency. *Normally the operating-frequency range of a PLL is restricted to the lock range.*

The discussion of the mechanical analogy did not provide numerical solutions for these four key parameters. We will derive such expressions in Sec. 3-2. The quantitative relationships between these four parameters are plotted in Fig. 3-8 for most practical cases. We can state in advance that the *hold range* $\Delta\omega_H$ is greater than the three remaining parameters. Furthermore we know that the pull-in range $\Delta\omega_P$ must be greater than the lock range $\Delta\omega_L$. The pull-in range $\Delta\omega_P$ is greater than the pull-out range $\Delta\omega_{PO}$ in most practical designs, so we get the simple inequality

$$\Delta\omega_L < \Delta\omega_{PO} < \Delta\omega_P < \Delta\omega_H$$

Keep in mind that the results gained by means of the mechanical analogy are valid for the *linear* PLL only. In the case of the *digital* PLL the situation is very much different, though simpler, as will be demonstrated in Chap. 4.

In many texts the term *capture range* can also be found. In most cases capture range is an alternative expression for lock range; sometimes it is also used to mean pull-in range. The differentiation between lock-in and pull-in ranges is not clearly established in some books, but we will maintain it throughout this text.[1]

## 3-2 LOCK-IN AND LOCK-OUT PROCESSES

In Sec. 3-1-3 the lock-in and lock-out performance of the PLL has been investigated qualitatively. We will now quantify these results.

### 3-2-1 The Hold Range

The hold range is the frequency range in which a PLL is able to maintain lock *statically*. The magnitude of the hold range is obtained by calculating that frequency offset at the reference input which causes a phase error $\theta_e$ of $\pi/2$. In this case we have

$$\omega_1 = \omega_0 + \Delta\omega_H \tag{3-21}$$

where $\Delta\omega_H$ is the hold range. For the phase signal $\theta_1(t)$ we get

$$\theta_1(t) = \Delta\omega_H \cdot t \tag{3-22}$$

The Laplace transform of the phase signal therefore becomes

$$\Theta_1(s) = \frac{\Delta\omega}{s^2} \tag{3-23}$$

The phase error can now be calculated according to Eq. (3-4):

$$\Theta_e(s) = \Theta_1(s)H_e(s) = \frac{\Delta\omega}{s^2}\frac{K_0K_dF(s)}{s + K_0K_dF(s)} \tag{3-24}$$

Using the final-value theorem of the Laplace transform (see Appendix F), we calculate the final phase error in the time domain:

$$\lim_{t\to\infty} \theta_e(t) = \lim_{s\to 0} s\Theta_e(s) = \frac{\Delta\omega}{K_0K_dF(0)} \tag{3-25}$$

Remember that the PLL network was linearized when the Laplace transform was introduced. Consequently Eq. (3-25) is valid for small values of $\theta_e$ only. For greater values of phase error we would have to write

$$\lim_{t\to\infty} \sin\theta_e(t) = \frac{\Delta\omega_H}{K_0K_dF(0)} \tag{3-26}$$

At the limit of the hold range, $\theta_e = \pi/2$ and $\sin\theta_e$ is exactly 1. Therefore we obtain for the hold range the expression

$$\Delta\omega_H = K_0K_dF(0) \tag{3-27}$$

Looking at the transfer functions of the loop filters in Table 2-2, we learn that for a passive filter $F(0)$ is just 1, whereas for an active loop filter $F(0) \to \infty$. Theoretically we obtain the following hold ranges of second-order PLLs:

Active loop filter:     $\Delta\omega_H \to \infty$

Passive loop filter:     $\Delta\omega_H = K_0K_d = $ loop gain $\tag{3-28}$

In practical applications the hold range of PLLs utilizing an active loop filter is as large as the frequency range covered by the VCO.

### 3-2-2 The Lock Range

The magnitude of the lock range can be obtained by a simple consideration. We assume that the PLL is initially not locked and that the reference frequency is $\omega_1 = \omega_0 + \Delta\omega$. According to Eq. (2-2) the phase detector will deliver an output signal given by

$$u_d(t) = K_d \{\sin (\Delta\omega \cdot t) + \sin [(2\omega_0 + \Delta\omega)t + \phi]\}$$

where $\phi$ is a zero phase. The second term can be discarded because it will be almost entirely filtered out by the loop filter. At the output of the loop filter there appears a signal $u_f(t)$ given by

$$u_f(t) \approx K_d |F(j\Delta\omega)| \sin (\Delta\omega t + \phi) \qquad (3\text{-}29)$$

This is an ac signal which causes a frequency modulation of the VCO. The peak frequency deviation is equal to $K_d K_0 |F(j\Delta\omega)|$.

In Fig. 3-9, the frequency $\omega_2$ of the VCO is plotted against time for two cases. In Fig. 3-9$a$ the peak frequency deviation is less than the offset $\Delta\omega$ between the reference frequency $\omega_1$ and the frequency of the VCO $\omega_2$.

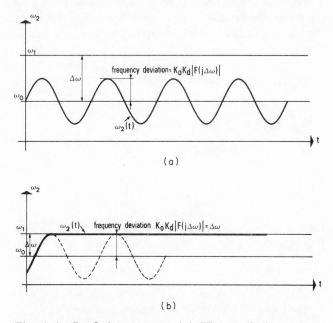

(a)

(b)

**Fig. 3-9   Lock-in process. ($a$) The peak frequency deviation is smaller than the offset $\Delta\omega$; therefore a fast lock-in process cannot take place. ($b$) The peak frequency deviation is exactly as large as the actual frequency offset, thus the PLL will become locked after a very short time.**

37

Hence a lock-in process will not take place, at least not instantaneously. Figure 3-9*b* shows a special case, however, where the peak frequency deviation is just as large as the frequency offset $\Delta\omega$. The frequency $\omega_2$ of the VCO output signal develops as shown by the solid line. When the frequency deviation is at its largest, $\omega_2$ exactly meets the value of the reference frequency $\omega_1$. Consequently the PLL locks within *one single-beat note* between the reference and output frequencies. This corresponds exactly to the lock-in process described in Sec. 3-1-2 by means of the mechanical analogy.

The condition for locking is therefore

$$K_0 K_d F(j\Delta\omega) \geq \Delta\omega \tag{3-30}$$

The lock range $\Delta\omega_L$ itself is consequently given by

$$\Delta\omega_L = K_0 K_d |F(j\Delta\omega_L)| \tag{3-31}$$

This is a nonlinear equation for $\Delta\omega_L$. Its solution becomes very simple, however, if an approximation is introduced for $|F(j\Delta\omega_L)|$. As shown in Table 2-2, various types of loop filters can be used for a PLL system; thus the solutions for $\Delta\omega_L$ are different for different types of filters. It can be shown by practical considerations that the lock range $\Delta\omega_L$ will always be considerably greater than the corner frequencies $1/\tau_1$ and $1/\tau_2$ of the loop filter; this is most easily seen when a PLL system is specified numerically. With this simplification, for type 1 and type 4 loop filters (Table 2-2):

$$|F(j\omega)| \approx \frac{1}{\omega\tau_1} \tag{3-32}$$

Substituting into Eq. (3-31), we get

$$\Delta\omega_L = \sqrt{\frac{K_0 K_d}{\tau_1}} \tag{3-33}$$

for type 1 and type 4 filters. Making use of the substitution in Eq. (3-6), this expression can be further simplified to

$$\Delta\omega_L \approx \omega_n \tag{3-34}$$

Equation (3-34) implies that the lock range is approximately as large as the natural frequency $\omega_n$, which is a very plausible result. When we plotted the Bode diagram of the phase-transfer function $H(j\omega)$ in Fig. 3-3, we saw that $\omega_n$ is the bandwidth within which a PLL is able to maintain phase tracking.

A similar result for the lock range $\Delta\omega_L$ is obtained for type 2 and type 3 loop filters. For these loop filters the gain becomes constant at higher frequencies, i.e., for the type 2 filter:

$$|F(j\omega)| \approx \frac{\tau_2}{\tau_1 + \tau_2} \qquad (3\text{-}35)$$

and for the type 3 filter:

$$|F(j\omega)| \approx \frac{\tau_2}{\tau_1} \qquad (3\text{-}36)$$

Making use of the substitutions in Eq. (3-6), we obtain for $\Delta\omega_L$ the simple expression

$$\Delta\omega_L \approx 2\zeta\omega_n \qquad (3\text{-}37)$$

for type 2 and type 3 filters.

Knowing the approximate size of the lock range we are certainly interested in having some indication of the lock-in time. According to Gardner[1] the lock-in time $T_L$ can be approximated by

$$T_L \approx \frac{1}{\omega_n} \qquad (3\text{-}38)$$

which is valid for any type of loop filter.

### 3-2-3 The Pull-In Range

If we consider Fig. 3-9a again, we see that the PLL has not yet locked and the peak frequency deviation is so small that the peak value of $\omega_2(t)$ never meets the reference frequency $\omega_1$. At a first glance one would think that the system would never become locked. We have to take into account, however, that the difference $\Delta\omega$ between reference frequency $\omega_1$ and output frequency $\omega_2(t)$ is not a constant; it is also varied by the frequency modulation of the VCO output signal. If the frequency $\omega_2(t)$ is modulated in the positive direction, the difference $\Delta\omega$ becomes smaller and reaches some minimum value $\Delta\omega_{min}$; if $\omega_2(t)$ is modulated in the negative direction, however, $\Delta\omega$ becomes greater and reaches some maximum value $\Delta\omega_{max}$. Because $\Delta\omega(t)$ is not a constant, the VCO frequency is modulated non-harmonically, that is, the duration of the half-period in which $\omega_2(t)$ is modulated in the positive direction becomes *longer* than that of the half-period in which $\omega_2(t)$ is modulated in the negative sense. This is shown graphically in Fig. 3-10. As a consequence the average frequency $\overline{\omega_2}$ of the VCO is now higher than it was without any modulation; i.e., the VCO frequency is pulled in the direction of the reference signal.

The asymmetry of the waveform $\omega_2(t)$ is greatly dependent on the value of the average offset $\Delta\omega$; the asymmetry becomes more marked as $\Delta\omega$ is decreased. If the average value of $\omega_2(t)$ is pulled somewhat in the

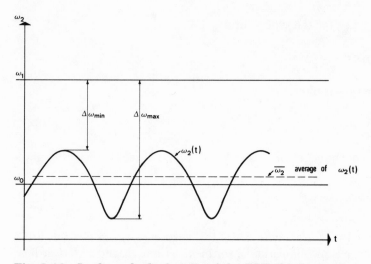

**Fig. 3-10** **In the unlocked state of the PLL the frequency modulation of the VCO output signal is nonharmonic. This causes the average value $\overline{\omega_2}$ of the VCO output frequency to be pulled in the direction of the reference frequency.**

direction of $\omega_1$ (which is assumed to be greater than $\overline{\omega_2}$), the asymmetry of the $\omega_2(t)$ waveform becomes stronger. This in turn causes $\overline{\omega_2}$ to be pulled even more in the positive direction. This process is regenerative under certain conditions, so that the output frequency $\omega_2$ finally reaches the reference frequency $\omega_1$. This phenomenon is called the *pull-in process* (Fig. 3-11). Mathematical analysis shows that a pull-in process occurs

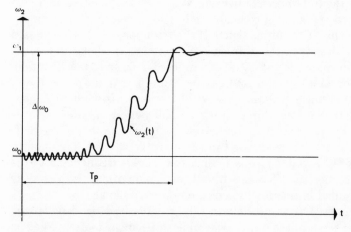

**Fig. 3-11** **The pull-in process.**

whenever the initial frequency offset $\Delta\omega$ is smaller than a critical value, the *pull-in range* $\Delta\omega_P$. If, on the other hand, the initial frequency offset $\Delta\omega$ is larger than $\Delta\omega_P$, a pull-in process does not take place because the pulling effect is not then regenerative. The mathematical treatment is quite cumbersome and is omitted here; it is given, however, in Appendix A. We shall consider only the final results of the analysis.[2]

For the pull-in range $\Delta\omega_P$ we get approximately

$$\Delta\omega_P \approx \frac{8}{\pi} \sqrt{\zeta\omega_n K_0 K_d - \omega_n^2} \tag{3-39}$$

For high-gain loops ($\omega_n \gg K_0 K_d$), this reduces to

$$\Delta\omega_P \approx \frac{8}{\pi} \sqrt{\zeta\omega_n K_0 K_d} \tag{3-40}$$

The pull-in process depicted in Fig. 3-11 can be easily explained by the mechanical analogy of Fig. 3-6. Initially the frequency offset $\Delta\omega_0$ is fairly large, and in the analogy the pendulum rotates at $\Delta\omega_0/2\pi$ revolutions per second. The angular velocity decelerates slowly, however, and the pendulum comes to rest after some time. The "pumping" of the instantaneous frequency $\omega_2(t)$ is very characteristic of this process and is easily explained by the nonharmonic rotation of the pendulum caused by the gravity of its mass $M$. In fact its angular velocity is greater at the lower "dead point" than at the higher one.

The duration of the pull-in process for all second-order PLLs is given approximately[1] by

$$T_P \approx \frac{\Delta\omega_0^2}{2\zeta\omega_n^3} \tag{3-41}$$

(Refer to Appendix A.) In this formula $\Delta\omega_0$ is the initial frequency offset $\omega_1 - \omega_2$ for $t = 0$. The quadratic and cubic terms in Eq. (3-41) show that the pull-in process is highly nonlinear. The pull-in time $T_P$ is normally much longer than the lock-in time $T_L$. This is demonstrated easily by a numerical example.

### Numerical Example

A second-order PLL having a type 2 (passive) loop filter is assumed to operate at a center frequency $f_0$ of 100 kHz. Its natural frequency $f_n = \omega_n/2\pi$ is 3 Hz; this is a very narrow-band system. The damping factor is chosen to be $\zeta = 0.7$. The loop gain $K_0 K_d$ is assumed to be $2\pi \cdot 1000$ s$^{-1}$. We shall now calculate the lock-in time $T_L$ and the pull-in time $T_P$ for an initial frequency offset $\Delta f$ of 30 Hz.

According to Eqs. (3-38) and (3-41) we get

$T_L$ = 5.3 ms

$T_P$ = 3.789 s

$T_P$ is practically 1000 times larger than $T_L$.

### 3-2-4 The Pull-Out Range

The pull-out range is by definition that frequency step which causes a lockout if applied to the reference input of the PLL. In the mechanical analogy of Fig. 3-6 the pull-out frequency corresponds to that weight that causes the pendulum to tip over if the weight is suddenly dropped onto the platform.

An exact calculation of the pull-out range is not possible for the linear PLL. However, simulations on an analog computer[1] have led to an approximation:

$$\Delta\omega_{PO} = 1.8\omega_n(\zeta + 1) \tag{3-42}$$

In most practical cases the pull-out range is between the lock range and the pull-in range

$$\Delta\omega_L < \Delta\omega_{PO} < \Delta\omega_P \tag{3-43}$$

If, in an FM system the PLL is pulled out by too large a frequency step, we can expect that PLL to come back to stable operation—by means of a relatively slow pull-in process—provided the frequency offset $\Delta\omega = \omega_1 - \omega_0$ is smaller than $\Delta\omega_P$. If the corresponding pull-in time is considered to be too long, the peak frequency deviation $\Delta\omega$ must be confined to the lock range $\Delta\omega_L$.

### 3-2-5 Determination of Lock-In and Lock-Out Processes by Computer Simulation

In Secs. 3-2-1 through 3-2-4 we found some quantitative results for the key parameters such as hold range, lock range, and so on. Most of these results were obtained by approximations. In this section we shall see that *exact* results for lock-in and lock-out processes can be found by solving the nonlinear differential equation of the PLL by means of a digital computer.

To investigate the tracking capability of a PLL, we assume that the system is initially locked and is disturbed at time $t = 0$ by applying a frequency step $\Delta\omega$ to its reference input. It is furthermore assumed that the loop has high gain, $\omega_n/K_0K_d = 0.1$. The damping factor is chosen to be 0.7.

Figure 3-12 plots the phase error $\theta_e(t)$ versus time $t$. As long as the frequency step $\Delta\omega$ is smaller than $2.5\omega_n$, the PLL stays locked at all times and settles at a relatively small phase error after some time (curves 1–3). At $\Delta\omega = 2.8\omega_n$ the PLL temporarily unlocks (curve 4), but becomes locked again. The final phase error is approximately $2\pi$, which means in the mechanical analogy (Fig. 3-6) that the pendulum tipped over once and came to rest thereafter. This corresponds to a pull-out process followed by a lock-in process.

At $\Delta\omega = 2.9\omega_n$ (curve 5) the PLL also unlocks, but the phase error increases more rapidly than in the previous case. It exceeds the value of $6\pi$, but the system locks again. In the mechanical analogy this corresponds to the case where the pendulum performs three revolutions before coming to rest.

If the frequency step $\Delta\omega$ is increased ($\Delta\omega = 3$, curve 6; and $\Delta\omega = 4$, curve 7), the system locks out forever. A pull-in process no longer takes place, and the phase error builds up to infinity.

In another computer simulation it was assumed that the same PLL was initially unlocked. Furthermore, the phase error at $t = 0$ was assumed to be $\pi$. The simulation investigated whether the PLL would lock if the reference frequency was offset by a constant $\Delta\omega$ from the center frequency $\omega_0$. In Fig. 3-13 the phase error $\theta_e(t)$ is plotted vs. time $t$ for different values of offset $\Delta\omega$. For the first three cases ($\Delta\omega = \omega_n$, $\Delta\omega = 1.4\omega_n$, and $\Delta\omega = 2\omega_n$, curves 1–3) the system locks within a short time period. For larger offsets ($\Delta\omega = 2.5\omega_n$ and $\Delta\omega = 3.0\omega_n$, curves 4 and 5) locking is no longer possible, and the PLL stays unlocked.

### 3-3 TRACKING PERFORMANCE OF LINEAR PLLs

In this section the tracking performance of the linear PLL is considered in detail. The following premises are assumed to be valid throughout this section:

1. A PLL is locked at $t = 0$.

2. A PLL is never subject to disturbances that could cause a lock-out process.

3. The noise signal at the reference input is negligible.

We can therefore make use of the linear model shown in Fig. 3-2.

In the following discussion, we will consider the transient response $\theta_e(t)$ of the second-order linear PLL for the most important modulations of the reference signal:

1. Phase modulation of the reference signal

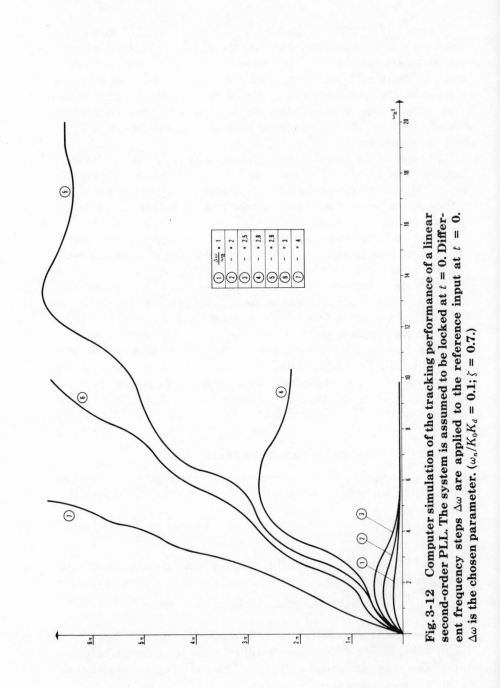

**Fig. 3-12** Computer simulation of the tracking performance of a linear second-order PLL. The system is assumed to be locked at $t = 0$. Different frequency steps $\Delta\omega$ are applied to the reference input at $t = 0$. $\Delta\omega$ is the chosen parameter. ($\omega_n/K_0K_d = 0.1$; $\zeta = 0.7$.)

| | $\frac{\Delta\omega}{\omega_n}$ = 1 |
|---|---|
| ① | $\frac{\Delta\omega}{\omega_n}$ = 1 |
| ② | " = 2 |
| ③ | " = 2.5 |
| ④ | " = 2.8 |
| ⑤ | " = 2.9 |
| ⑥ | " = 3 |
| ⑦ | " = 4 |

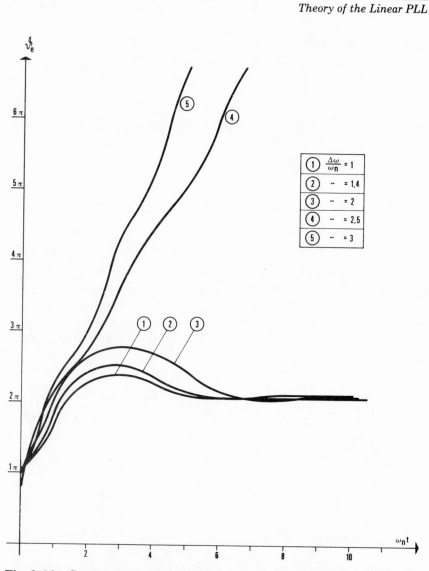

**Fig. 3-13** Computer simulation of the acquisition performance of a linear second-order PLL. The system is assumed to be unlocked at $t = 0$, the initial phase error $\theta_e (0)$ being $\pi$. The offset $\Delta\omega = \omega_1 - \omega_0$ is the chosen parameter. ($\omega_n/K_0K_d = 0.1$; $\zeta = 0.7$.)

2. Frequency modulation of the reference signal by a sine wave

3. Frequency modulation of the reference signal by a square-wave signal (frequency shift keying, or FSK)

4. Frequency modulation of the reference signal by a ramp signal

### 3-3-1 Phase Modulation

Assume that the reference signal is subject to a phase step at $t = 0$. For the phase signal we have

$$\theta_1(t) = u(t) \, \Delta\Phi \tag{3-44}$$

where $\Delta\Phi$ is the amplitude of the phase step (refer also to Fig. 3-1$a$. For the Laplace transform of $\theta_1(t)$ we have

$$\Theta_1(s) = \frac{\Delta\Phi}{s} \tag{3-45}$$

According to Eq. (3-4) the Laplace transform of the error phase $\theta_e(t)$ becomes

$$\Theta_e(s) = H_e(s)\Theta_1(s) \tag{3-46}$$

The error phase $\theta_e(t)$ is obtained by performing the inverse Laplace transform on Eq. (3-46). This is done in detail in Appendix B. Here we shall consider only the final result. In Fig. 3-14, the phase error $\theta_e(t)$ is plotted against time for different values of the damping factor $\zeta$. For simplicity it is assumed that the loop is a high-gain loop. Hence $\omega_n/K_dK_0$ is set at zero.

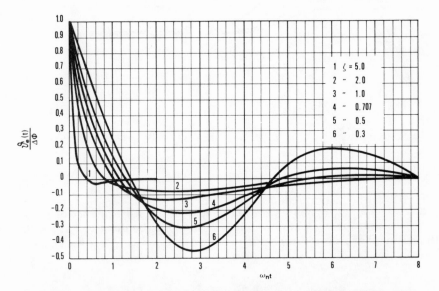

**Fig. 3-14  Transient response of a linear second-order PLL to a phase step $\Delta\Phi$ applied at $t = 0$. The PLL is assumed to be a high-gain loop. (*Adapted from Gardner[1] with permission.*)**

It is seen from Fig. 3-14 that for a second-order PLL $\theta_e(t)$ never exceeds the initial phase step $\Delta\Phi$. By means of the mechanical analogy of Fig. 3-6 we can state that the system stays locked for phase steps $\Delta\Phi$ smaller than $\pi$. This corresponds to an impulse applied to the platform which causes the pendulum to swing to its culmination point without tipping over.

In practical applications the peak phase step will be limited to about $\pi/2$ to prevent unlocking the system as a result of superimposed noise signals.

As seen from Fig. 3-14 the phase error always decays to zero after some time. This is easily proved by applying Eqs. (3-4) and (3-46). For $\Theta_e(s)$ we can write

$$\Theta_e(s) = \frac{\Delta\Phi}{s} \cdot \frac{s}{s + K_0 K_d F(s)} \tag{3-47}$$

Using the final-value theorem of the Laplace transform, we get

$$\lim_{t\to\infty} \theta_e = \lim_{s\to 0} s\Theta_e(s) = s\frac{\Delta\Phi}{s} \cdot \frac{s}{s + K_0 K_d F(0)} = 0 \tag{3-48}$$

In control theory $\theta_e(\infty)$ is also called the *position error*. The position error is therefore always zero for second-order PLLs.

### 3-3-2 Frequency Modulation

3-3-2-1 Frequency Modulation by a Sine-Wave Signal   In the mechanical analogy (Fig. 3-6), the frequency modulation by a sine wave corresponds to a harmonic excitation of the weighing platform, such as by pressing it down periodically with the modulating frequency. The pendulum then starts oscillating around its quiescent position. Its amplitude will be largest if it is excited with its natural frequency $\omega_n'$. In this case, for the reference signal we write

$$\omega_1(t) = \omega_0 + \Delta\omega \sin \omega_m t \tag{3-49}$$

Here $\Delta\omega$ is the peak frequency deviation and $\omega_m$ is the angular frequency of the modulating signal.

By integrating Eq. (3-49) over time $t$, we obtain the phase signal $\theta_1(t)$:

$$\theta_1(t) = -\frac{\Delta\omega}{\omega_m} \cos \omega_m t \tag{3-50}$$

where $\theta_1(t)$ is a sine-wave signal having the amplitude $\Delta\omega/\omega_m$. The peak amplitude of the phase error $\hat{\theta}_e(t)$ is therefore easily obtained by multiplying $\Delta\omega/\omega_m$ by the magnitude $|H_e(j\omega)|$ of the error-transfer function at the modulating frequency:

$$\hat{\theta}_e(j\omega_m) = \frac{\Delta\omega}{\omega_m} \, |H_e(j\omega_m)| \qquad (3\text{-}51)$$

Figure 3-15 shows the peak phase error $\hat{\theta}_e$ as a function of the modulating frequency $\omega_m$. The damping factor $\zeta$ is the chosen parameter. As expected, the curve has a marked peak at $\omega_n = \omega_m$; the amplitude increases with decreasing $\zeta$.

**3-2-2 Frequency Modulation by a Square-Wave Signal** For FSK the modulating signal is a square wave. We are interested in knowing the PLL's transient response to a single frequency step only; thus we can write for the reference frequency:

$$\omega_1 = \omega_0 + \Delta\omega \, u(t) \qquad (3\text{-}52)$$

(Refer also to Fig. 3-1$b$.) For $\theta_1(t)$ we have, consequently,

$$\theta_1(t) = \Delta\omega \, t \qquad (3\text{-}53)$$

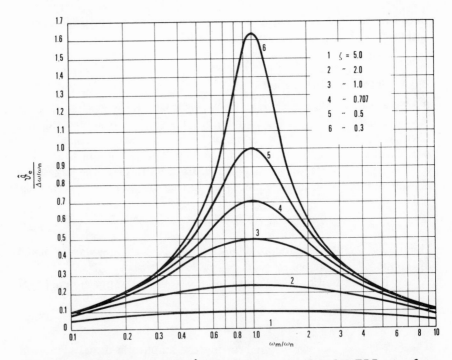

**Fig. 3-15** Peak phase error $\hat{\theta}_e$ of a linear second-order PLL at a frequency modulation of the reference signal by a sine wave. *(Adapted from Gardner[1] with permission.)*

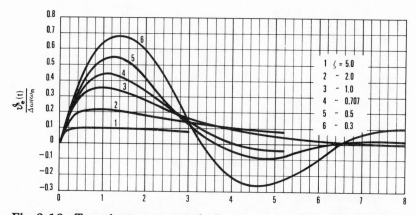

**Fig. 3-16** Transient response of a linear second-order PLL to a frequency step $\Delta\omega$ applied to its reference input at $t = 0$. *(Adapted from Gardner[1] with permission.)*

and $\Theta_1(s)$ becomes

$$\Theta_1(s) = \frac{\Delta\omega}{s^2} \qquad (3\text{-}54)$$

Applying Eq.(3-4). we get

$$\Theta_e(s) = \Theta_1(s) \cdot H_e(s) \qquad (3\text{-}55)$$

The phase error $\theta_e(t)$ is found by performing the inverse Laplace transform on Eq. (3-55). (This derivation is given in Appendix C.) Here only the final result is plotted in Fig. 3-16. It is assumed that the PLL is a high-gain loop; thus $\omega_n/K_0K_d$ can be set to zero. The damping factor $\zeta$ was the chosen parameter. Figure 3-16 shows that the transient response $\theta_e(t)$ is oscillatory for small values of $\zeta$. For $\zeta > 1$, however, the response becomes aperiodic. The transient dies out fastest if we select for $\zeta$ the value

$$\zeta = \zeta_{\text{opt}} = \frac{\sqrt{2}}{2} \approx 0.7 \qquad (3\text{-}56)$$

(This corresponds to the dynamic response of a second-order Butterworth low-pass filter.)

We will now calculate the steady-state phase error $\theta_e(\infty)$ for the case of FSK. Here the exciting function $\theta_1(t)$ is a ramp [Eq. (3-52)]; in control theory the corresponding state error is called the *velocity error*.[3] Applying the final-value theorem of the Laplace transform[4] to Eq. (3-55), we get

$$\theta_e(\infty) = \frac{\Delta\omega}{K_0K_dF(0)} \qquad (3\text{-}57)$$

Now we remember that $F(0)$ is unity for passive loop filters, but is theoretically infinite for active type 3 and type 4 loop filters (Table 2-2). Consequently we obtain for a passive loop filter:

$$\theta_e(\infty) = \frac{\Delta\omega}{K_0 K_d} \tag{3-58a}$$

and for an active loop filter:

$$\theta_e(\infty) = 0 \tag{3-58b}$$

Let us now look at a numerical example.

### Numerical Example

A second-order PLL (high-gain loop) operates at a center frequency $f_0$ of 100 kHz. The damping factor chosen is $\zeta = 0.7$. The natural frequency $f_n = \omega_n/2\pi$ is 10 kHz. The reference signal is subject to FSK. How large will the maximum phase error be at a step frequency deviation of 20 kHz? Will the system still be stable in this case? If yes, what is the size of the frequency step causing the PLL to unlock?

Our numerical data are

$$\omega_n = 2\pi f_n = 6.28 \times 10^4 \text{ s}^{-1}$$

$$\zeta = 0.7$$

$$\Delta\omega = 2\pi \, \Delta f = 1.25 \times 10^5 \text{ s}^{-1}$$

From Fig. 3-16 we read that the peak value of

$$\frac{\theta_e}{\Delta\omega/\omega_n}$$

is 0.46 for $\zeta = 0.7$. Hence the peak phase error becomes

$$\theta_{e \text{ max}} = \frac{\Delta\omega}{\omega_n} \cdot 0.46 = 0.92 \text{ rad } (\cong 52.7°)$$

This is positively less than $\pi$, hence the system remains locked. The pull-out frequency cannot be read from Fig. 3-16 because the curves were derived by the linearized PLL model. If we were trying to find which frequency step $\Delta\omega$ from Fig. 3-16 would generate a peak phase error of $\pi$, we would get the faulty result $\Delta f = \Delta\omega/2\pi = 68$ kHz. To find the correct answer we must use the formula for the pull-out range $\Delta\omega_{\text{PO}}$ (Eq. 3-42). This yields

$$\Delta\omega_{\text{PO}} = 1.8\omega_n(\zeta + 1)$$

or

$$\Delta f_{\text{PO}} = 1.8 f_n(0.7 + 1) = 3.1 f_n = 31 \text{ kHz}$$

The frequency step causing the PLL to unlock is 31 kHz, not 68 kHz.

**3-3-2-3 Frequency Modulation by a Ramp Signal**   If the reference input is frequency modulated by a ramp function, the reference frequency $\omega_1$ is built up linearly with time (refer also to Fig. 3-1c). When simulating this by using the mathematical pendulum in Fig. 3-6, we would have to pile some convenient material on the platform. It is clear that the pendulum will become unstable as soon as the weight on the platform becomes larger than the weight causing a steady-state error of more than $\pi/2$. But the system can become unstable earlier if the feed rate of the mechanical analogy is too great. This will be shown by the following derivation.

For $\omega_1(t)$ we have

$$\omega_1(t) = \omega_0 + \overset{.}{\Delta\omega}\, t \tag{3-59}$$

where $\overset{.}{\Delta\omega}$ is again the rate of change of the reference frequency. Therefore the phase signal $\theta_1(t)$ is given by

$$\theta_1(t) = \overset{.}{\Delta\omega}\, \frac{t^2}{2} \tag{3-60}$$

In the complex frequency domain we then have

$$\Theta_1(s) = \frac{\overset{.}{\Delta\omega}}{s^3} \tag{3-61}$$

By applying Eq. (3-4), we calculate the phase error $\theta_e(t)$ as follows:

$$\Theta_e(s) = H_e(s) \cdot \Theta_1(s) = \frac{\overset{.}{\Delta\omega}}{s^3} \cdot \frac{s}{s + K_0 K_d F(s)} \tag{3-62}$$

Using the final-value theorem of the Laplace transform, we get

$$\theta_e(\infty) = \lim_{s \to 0} s\, \frac{\overset{.}{\Delta\omega}}{s^3} \cdot \frac{s}{s + K_0 K_d F(s)} \tag{3-63}$$

In control theory this error is called the *acceleration error*. The acceleration error is by definition the steady-state error on an exciting function which is a quadratic function of time.

If we assume a passive loop filter, $F(0)$ is unity. Introducing this into Eq. (3-63) yields

$$\theta_e(\infty) = \infty$$

This is a strange result, but it is nevertheless true because the phase error really approaches infinity after a long enough time. In fact, the PLL unlocks if the hold range $\Delta\omega_H$ is exceeded. In the case of the passive loop filter, the final-value theorem does not deliver a useful solution for the maximum allowable rate of change of the reference frequency $\overset{.}{\Delta\omega}$.

The situation is more favorable for the active loop filter. Introducing the transfer function $F(s)$ for the active loop filter (refer to Table 2-2) into Eq. (3-63), we get

$$\theta_e(\infty) = \frac{\dot{\Delta\omega}\,\tau_1}{K_0 K_d}$$

Making use of the substitutions in Eq. (3-6), we obtain

$$\theta_e(\infty) = \frac{\dot{\Delta\omega}}{\omega_n^2} \tag{3-64}$$

This is a remarkable result. It tells us that the phase error remains finite if an active loop filter is used. According to Eq. (3-64) the error remains finite even after an arbitrarily long time interval. This is explained by the fact that the hold range is also theoretically infinite if an active loop filter is used. In practice this cannot be true, of course, because the operating frequency range of the VCO is limited. Equation (3-64) is therefore valid only within the operating frequency range of the PLL.

Remember that Eq. (3-64) was obtained by using the linearized model of the PLL; hence it is valid for small values of $\theta_e(\infty)$ only. For larger values of $\theta_e$, the correct expression is

$$\sin\theta_e(\infty) = \frac{\dot{\Delta\omega}}{\omega_n^2} \tag{3-65}$$

Because the sine function cannot exceed unity, the maximum rate of change of the reference frequency that does not cause lock-out is

$$\dot{\Delta\omega}_{max} = \omega_n^2 \tag{3-66}$$

This result has two consequences:

1. If the reference frequency is swept at a rate larger than $\omega_n^2$, the system will unlock.

2. If the system is initially unlocked, it cannot become locked if the reference frequency is simultaneously swept at a rate larger than $\omega_n^2$. (Refer also to Sec. 3-4-3.)

Practical experience with PLLs has shown that the formula $\dot{\Delta\omega}_{max} = \omega_n^2$ is a theoretical limit which is normally not practicable. If the reference frequency is swept in the presence of noise, the rate at which an initially unlocked PLL can become locked is markedly less than $\omega_n^2$.[1,5,6] A more practical design limit[1] for $\dot{\Delta\omega}$ is considered to be

$$\dot{\Delta\omega}_{max} < \tfrac{1}{2}\omega_n^2 \tag{3-67}$$

The acceleration error can be made zero if the loop filter has a double pole at $s = 0$. [Refer to Eq. (3-64).] This corresponds to two integrators connected in cascade; the overall system would be a third-order PLL.

## 3-4 NOISE IN LINEAR PLL SYSTEMS

### 3-4-1 Simplified Noise Theory

The theory of noise in PLLs is very cumbersome. Exact solutions for noise performance have been derived for first-order PLLs only[7]; for second-order PLLs, computer simulations have been made which provided approximate results.[2] We will deal here only with some final results of the noise theory and discuss a few key parameters which have a marked effect on the practical design of a PLL system. The interested reader is pointed toward the specialized noise literature.[2,7]

Noise performance is most easily considered by means of the well-known mechanical analogy shown in Fig. 3-17. The noise theory reveals that a noise signal $u_{n1}(t)$ superimposed on the (noiseless) reference signal $u_1(t)$ is characterized in the model by two stochastic perpendicular forces $f_1(t)$ and $f_2(t)$; $f_1(t)$ and $f_2(t)$ are totally uncorrelated. These disturbing forces are likely to have an adverse effect on the acquisition and tracking performances of the PLL.

For a low noise amplitude the pendulum in Fig. 3-17 jitters around its quiescent position only slightly; the system remains stable. If the noise amplitude is made larger, the pendulum will tip over from time to time on a larger noise peak, but nevertheless it stays near its quiescent position most of the time.

The probability that the pendulum will be found near its point of

**Fig. 3-17 Mechanical model for a PLL having a noisy reference signal; $f_1$ and $f_2$ are stochastic forces (disturbances) as explained in text.**

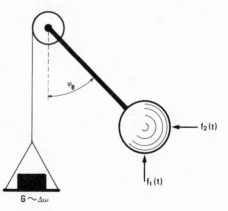

equilibrium is much higher than the probability of its being seen at a completely different location. If the noise amplitude is increased further, however, no stable operating point will be established. The pendulum is now jittering around irregularly, and the probability that it will be found in the sector $\phi_e \cdots \phi_e + \Delta\phi_e$ (where $\phi_e$ is the deflection angle) at a time $t$ is identical for all angles $\phi_e$ in the range $0-2\pi$. Now, of course, the PLL can no longer become locked.

For the designer of a PLL system the following questions are of interest:

1. How large must the signal-to-noise ratio (SNR) at the reference input be to enable safe acquisition of the PLL?

2. Provided the SNR is large enough to enable the lock-in process, how often on the average then does the PLL temporarily unlock (how many times per second)?

These questions are answered by the noise theory (at least to a useful approximation). The most important parameters describing a noise signal are explained with the aid of Fig. 3-18. The corresponding power spectra are shown in Fig. 3-19. All noise signals considered here are defined as white noise, i.e., the power spectrum is flat; this means that the same noise power $\Delta P$ is contained in every frequency interval $\Delta f$ ($dP/df$ is a constant). (Refer to Fig. 3-19$a$.) The bandwidth of the noise spectrum is limited to $B_i$, mostly by a prefilter. Even without a prefilter the noise bandwidth would be limited for other reasons, for the total noise power cannot be infinite. In most practical applications the prefilter is a bandpass or low-pass filter; $B_i$ is the bandwidth of such a filter. The noise signal is assumed to have a total noise power $P_n$ (in watts). The reference (input) signal of the PLL is assumed to be narrow-band, so its spectrum in Fig. 3-19$a$ may be characterized by one single line. Its power is defined by $P_s$, where $P_s$ is normally given by

$$P_s = \frac{U_1^2 \, (\text{rms})}{R_1}$$

where $R_1$ is the input impedance of the PLL and $U_1$ (rms) is the rms value of the input signal. The prefilter is assumed to have an ideal transfer function; i.e., its gain is unity inside the passband and zero outside of it.

The spectral density of the noise signal is defined as

$$W_i = \frac{P_n}{B_i} \qquad \text{W Hz}^{-1} \tag{3-68}$$

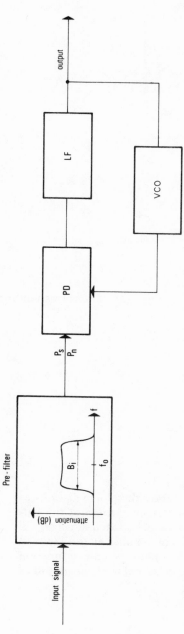

**Fig. 3-18  Noise suppression.** ($f_0 = \omega_0/2\pi$; $B_i$ is the prefilter bandwidth; $P_s$ is the effective signal power output at the PLL input; $P_n$ is the noise power output at the PLL input; the rest of the symbols are used in Fig. 1-1$a$.)

55

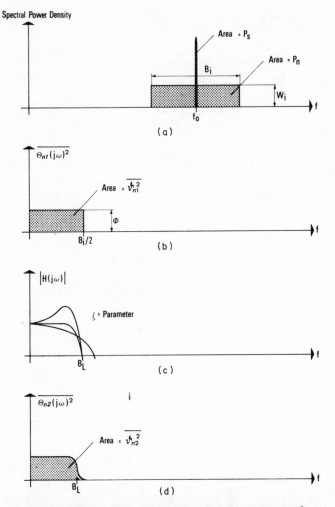

Fig. 3-19 Method of calculating the phase jitter $\theta_{n2}^2$ at the output of the PLL. (**a**) Power spectra of the reference signal $u_1(t)$ and the superimposed noise signal $u_n(t)$. (**b**) Spectrum of the phase noise at the input of the PLL. (**c**) Bode plot of the phase-transfer function $H(j\omega)$. (**d**) Spectrum of the phase noise at the output of the PLL.

(Refer to Fig. 3-19$a$.) It is postulated now that the reference signal $u_1(t)$ is a sine wave. If a noise signal is added to $u_1(t)$, the zero crossings of the resulting signal are displaced back or forward depending on the instantaneous polarity of the noise signal. A so-called phase jitter is generated, designated $\phi_{n1}(t)$ (phase noise).

It can be shown[1] that the rms value of the phase jitter $\overline{\theta_{n1}^2}$ (or, more exactly, the square of the rms phase noise) is given by the simple expression

$$\overline{\theta_{n1}^2} = \frac{P_n}{2P_s} \tag{3-69}$$

We now define the SNR at the input of the PLL as

$$(\text{SNR})_i = \frac{P_s}{P_n} \tag{3-70}$$

For the phase jitter at the input of the PLL we get the simple relation

$$\overline{\theta_{n1}^2} = \frac{1}{2(\text{SNR})_i} \qquad \text{rad}^2 \tag{3-71}$$

that is, the square of the rms value of the phase jitter is inversely proportional to the SNR at the input of the PLL. According to the theory of noise[1] the phase jitter $\overline{\theta_{n1}^2}(t)$ can also be represented as a frequency spectrum. This spectrum reaches from zero to $B_i/2$ and is shown in Fig. 3-19$b$. The spectral density of the phase jitter is denoted as $\Theta_{n1}(j\omega)$. The square of the spectral density $\overline{\Theta_{n1}^2(j\omega)}$ is then

$$\overline{\Theta_{n1}^2(j\omega)} = \Phi = \frac{\overline{\theta_{n1}^2}}{B_i/2} \qquad \text{rad}^2\,\text{Hz}^{-1} \tag{3-72}$$

Because the frequency spectrum of phase noise $\Theta_{n1}(j\omega)$ is known at the input of the PLL, we are able to calculate the frequency spectrum at its output, too. The rms phase jitter at the output is defined as $\sqrt{\overline{\theta_{n2}^2}(t)}$, its frequency spectrum as $\Theta_{n2}(j\omega)$. Using Eq. (3-3) we can write

$$\sqrt{\overline{\Theta_{n2}^2(j\omega)}} = |H(j\omega)| \ \sqrt{\overline{\Theta_{n1}^2(j\omega)}} \tag{3-73}$$
$$= |H(j\omega)| \sqrt{\Phi}$$

This relationship is demonstrated in Fig. 3-19. Fig. 3-19$c$ shows the Bode diagram $|H(j\omega)|$, and Fig. 3-19$d$ shows the output-phase noise spectrum obtained by the multiplication of the curves shown in Fig. 3-19$b$ and $c$.

We are effectively looking for the rms value of the phase noise $\overline{\theta_{n2}^2}(t)$ at the PLL's output. The phase noise is calculated by integrating

$\Theta_{n2}(j\omega)$ over the bandwidth of the PLL

$$\overline{\theta^2_{n2}} = \int_0^\infty \overline{\Theta^2_{n2}}\,(j2\pi f)\,df \tag{3-74}$$

where $2\pi f = \omega$. $\overline{\theta^2_{n2}}$ is the area under the curve in Fig. 3-19$d$. Making use of Eqs. (3-73) and (3-74), we get

$$\overline{\theta^2_{n2}} = \int_0^\infty \Phi\,|H(j2\pi f)|^2\,df \tag{3-75}$$

$$= \frac{\Phi}{2} \int_0^\infty |H(j\omega)|^2\,d\omega$$

The integral $\displaystyle\int_0^\infty |H(j2\pi f)|^2\,df$ is called the *noise bandwidth* $B_L$,

$$B_L = \int_0^\infty |H(j2\pi f)|^2\,df \tag{3-76}$$

The solution of this integral reads[1]

$$B_L = \frac{\omega_n}{2}\left(\zeta + \frac{1}{4\zeta}\right) \tag{3-77}$$

Thus $B_L$ is proportional to the natural frequency $\omega_n$ of the PLL; furthermore, it depends on the damping factor $\zeta$. Figure 3-20 is a plot of $B_L/\omega_n$ vs. $\zeta$, and $B_L$ has a minimum at $\zeta = 0.5$. In this case we have

$$B_L = B_{L\,\text{min}} = \frac{\omega_n}{2} \tag{3-78}$$

In Sec. 3-3-2-2 we showed that the transient response of the PLL is best at $\zeta \approx 0.7$. Because the function $B_L(\zeta)$ is fairly flat in the neighborhood of $\zeta = 0.5$, the choice of $\zeta = 0.7$ does not noticeably worsen the noise performance. For $\zeta = 0.7$, $B_L$ is $0.53\omega_n$ instead of the minimum value $0.5\omega_n$. For this reason $\zeta = 0.7$ is chosen for most applications. For the output phase jitter $\overline{\theta^2_{n2}}(t)$ we can now write

$$\overline{\theta^2_{n2}} = \Phi B_L \tag{3-79}$$

where $\Phi$ is already known from Eq. (3-72). Combining Eqs. (3-69), (3-72), and (3-79), we obtain

$$\overline{\theta^2_{n2}} = \frac{P_n}{P_s} \cdot \frac{B_i}{B_L} \tag{3-80}$$

We saw in Eq. (3-71) that the phase jitter at the input of the PLL is inversely proportional to the (SNR)$_i$. By analogy we can also define a sig-

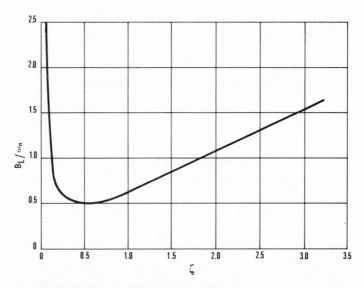

**Fig. 3-20  Noise bandwidth of a linear second-order PLL as a function of the damping factor $\zeta$.** *(Adapted from Gardner[1] with permission.)*

nal-to-noise ratio at the output, which will be denoted by $(SNR)_L$ (SNR of the loop). We define this analogy in Eq. (3-81),

$$\overline{\theta_{n2}^2} = \frac{1}{2(SNR)_L} \qquad \text{rad}^2 \tag{3-81}$$

Comparing Eqs. (3-80) and (3-81), we get

$$(SNR)_L = \frac{P_s}{P_n} \cdot \frac{B_i}{2B_L} = (SNR)_i \cdot \frac{B_i}{2B_L} \tag{3-82}$$

Equation (3-82) says that the PLL *improves* the SNR of the input signal by a factor of $B_i/2B_L$. The narrower the noise bandwidth $B_L$ of the PLL, the greater the improvement.

All this sounds very theoretical. Let us therefore look at some numerical data. In radio and television the SNR is used to specify the quality of information transmission. For a stereo receiver a minimum SNR of 20 dB is considered a fair design goal.* The same holds true for PLLs. Practical

---

*The SNR of a signal can be specified either numerically or in decibels. The $(SNR)_{dB}$ is calculated from

$$(SNR)_{dB} = 10 \log_{10} \frac{P_s}{P_n} = 10 \log_{10} \frac{U_s^2 \, (\text{rms})}{U_n^2 \, (\text{rms})} = 20 \log_{10} \frac{U_s \, (\text{rms})}{U_n \, (\text{rms})}$$

where $U_s$ (rms) and $U_n$ (rms) are the rms values of signal and noise, respectively.

experiments performed with second-order PLLs have demonstrated some very useful results.[1]

1.  For $(\text{SNR})_L = 1$ (0 dB), a lock-in process will not occur because the output phase noise $\sqrt{\overline{\theta_{n2}^2}}$ is excessive.

2.  At $(\text{SNR})_L = 2$ (3 dB), lock-in is eventually possible.

3.  For $(\text{SNR})_L = 4$ (6 dB), stable operation is generally possible.

In quantitative terms according to Eq. (3-81), for $(\text{SNR})_L = 4$ the output phase noise becomes

$$\overline{\theta_{n2}^2} = \frac{1}{2 \cdot 4} = \frac{1}{8} \quad \text{rad}^2$$

Hence the rms value $\sqrt{\overline{\theta_{n2}^2}}$ becomes

$$\sqrt{\overline{\theta_{n2}^2}} = \sqrt{\tfrac{1}{8}} = 0.353 \text{ rad} \ (\cong 20°)$$

Since the rms value of phase jitter is about 20°, the value of 180° (limit of dynamic stability) is rarely exceeded. Consequently the PLL does not unlock frequently. At $(\text{SNR})_L = 1$, however, the effective value of output phase jitter would be as large as 40°, and the dynamic limit of stability of $\pi$ would be exceeded on every major noise peak, thus making stable operation impossible.

As a rule of thumb,

$$(\text{SNR})_L \geq 4 \ (\cong 6 \text{ dB}) \tag{3-83}$$

is a convenient design goal.

The designer of practical PLL circuits is vitally interested in how often on the average a system will temporarily unlock. The probability of unlocking is decreased with increasing $(\text{SNR})_L$. We now define $T_{av}$ to be the average time interval between two lockouts. For example, if $T_{av} = 100$ ms, the PLL unlocks on an average 10 times per second. For second-order PLLs $T_{av}$ has been found experimentally as a function of $(\text{SNR})_L$.[8] The resulting curve is plotted in Fig. 3-21.

To illustrate the theory, let us calculate a numerical example.

### Numerical Example

A second-order PLL is assumed to have the following specifications:

$$f_n = 10 \text{ kHz}$$

$$\omega_n = 62.8 \times 10^3 \text{ s}^{-1}$$

$$(\text{SNR})_L = 1.5$$

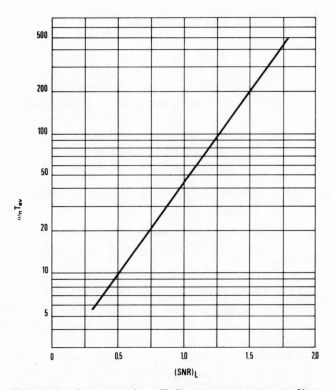

**Fig. 3-21** Average time $T_{av}$ between two succeeding unlocking events plotted as a function of the $(SNR)_L$ at the output of the PLL. $T_{av}$ is normalized to the natural frequency $\omega_n$ of the PLL. Note that this result is valid only for linear second-order PLLs. *(Adapted from Gardner[1] with permission.)*

From Fig. 3-21 we read $\omega_n T_{av} = 200$. Consequently $T_{av} \approx 3$ ms. This means that the PLL unlocks about 300 times per second, which looks quite bad. However, the lock-in time $T_L$ is found from Eq. (3-38) to be $T_L \approx 1/\omega_n = 15.9$ $\mu$s; that is, the PLL nevertheless is locked for 99.5 percent of the total time.

### A Summary of Noise Theory

Although the noise theory of the PLL is difficult, for practical design purposes it suffices to remember some rules of thumb:

1. Stable operation of the PLL is possible if $(SNR)_L$ is approximately 4.

2. $(SNR)_L$ is calculated from

$$(SNR)_L = \frac{P_s}{P_n} \cdot \frac{B_i}{2B_L}$$

where $P_s$ = signal power at the reference input

$P_n$ = noise power at the reference input

$B_i$ = bandwidth of the prefilter, if any; if no prefilter is used, $B_i$ is the bandwidth of the signal source (e.g., antenna, repeater)

$B_L$ = noise bandwidth of the PLL

3. The noise bandwidth $B_L$ is a function of $\omega_n$ and $\zeta$. For $\zeta = 0.7$, $B_L = 0.53\omega_n$.

4. The average time interval $T_{av}$ between two unlocking events gets longer as $(SNR)_L$ increases.

### 3-4-2 The Lock-In Process in the Presence of Noise

In a PLL application, if noise is superimposed on the reference signal, conflicting demands could arise if the designer tries to specify the key parameters of the system, such as lock range $\Delta\omega_L$, noise bandwidth $B_L$, and so on. If it is used as a communication receiver, the PLL is required to lock onto an input signal which is known to be within the frequency range $\omega_0 - \Delta\omega$ to $\omega_0 + \Delta\omega$. In order to lock onto this signal, the lock range $\Delta\omega_L$ should be at least as large as $\Delta\omega$:

$$\Delta\omega_L \geq \Delta\omega$$

When a suitable value has been chosen for the lock range $\Delta\omega_L$, the designer will calculate the natural frequency $\omega_n$ using Eq. (3-34) or Eq. (3-37) according to the type of loop filter specified. Assuming that a type 2 or type 3 loop filter (Table 2-2) has been selected and a damping factor of $\zeta = 0.7$ is used, we get

$$\Delta\omega_L = 1.4\omega_n$$

With $\omega_n$ known, the noise bandwidth $B_L$ is known, too. According to Fig. 3-20, $B_L$ becomes

$$B_L = 0.53\omega_n$$

The designer will now have to check whether or not this noise bandwidth yields a satisfactory $(SNR)_L$. From the given data it will be possible for the designer to specify the signal power $P_s$, the noise power $P_n$, and the prefilter bandwidth $B_i$. (Refer to Sec. 3-4-1.) From these data $(SNR)_L$ can then be calculated by using Eq. (3-82):

$$(SNR)_L = \frac{P_s}{P_n} \cdot \frac{B_i}{2B_L}$$

If it turns out that $(SNR)_L$ is at least approximately 4, we can be satisfied. If the reverse is true, however, we are in the dilemma of reducing the noise bandwidth $B_L$ accordingly. But this also reduces $\omega_n$, and consequently the lock range $\Delta\omega_L$, and thus the signal may possibly no longer be captured by the PLL. The contradiction seems unsurmountable, but the problem can be circumvented by the special pull-in techniques to be discussed in the next section.

### 3-4-3 Pull-In Techniques for Noisy Signals

In this section we will summarize the most popular pull-in techniques. A special pull-in procedure becomes mandatory if the noise bandwidth of a PLL system must be made so narrow that the signal may possibly not be captured at all.

a. The Sweep Technique   If this procedure is chosen, the noise bandwidth $B_L$ is made so small that the SNR of the loop $(SNR)_L$ is sufficiently large to provide stable operation. As a consequence the lock range $\Delta\omega_L$ might become smaller than the frequency interval $\Delta\omega$ within which the input signal is expected to be. To solve the locking problem, the center frequency of the VCO is swept by means of a sweeping signal $u_{sweep}$ (see Fig. 3-22) over the frequency range of interest. Of course the sweep rate $\dot{\Delta\omega}$ must be held within the limits specified by Eq. (3-67); otherwise the PLL could not become locked. (In Sec. 3-3-2-3 we considered the case where the center frequency $\omega_0$ was constant and the reference frequency $\omega_1$ was swept; in the case considered *here* the reference frequency is assumed to be constant, whereas the center frequency is swept. For the PLL both situations are equivalent, because the frequency offset $\Delta\omega = \omega_1 - \omega_0$ is the only parameter of importance.)

As shown in the block diagram of Fig. 3-22, the center frequency of the VCO can be tuned by the signal applied to its sweep input. A linear sawtooth signal generated by a simple $RC$ integrator is used as the sweep signal. Assume the PLL has not yet locked. The integrator is in its RUN mode, and hence the sweep signal builds up in the positive direction. As soon as the frequency of the VCO approaches the frequency of the input signal, the PLL suddenly locks. The sweep signal should now be frozen at its present value (otherwise the VCO frequency would run away). This is realized by throwing the analog switch in Fig. 3-22 to the HOLD position.

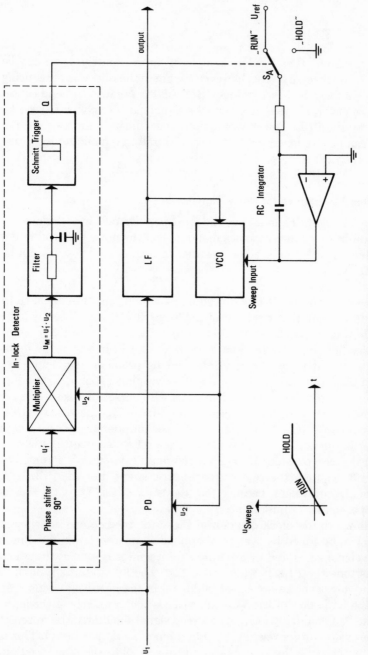

**Fig. 3-22  Simplified block diagram of a PLL using the sweep technique for the acquisition of signals buried in noise. ($S_A$ = analog switch.)**

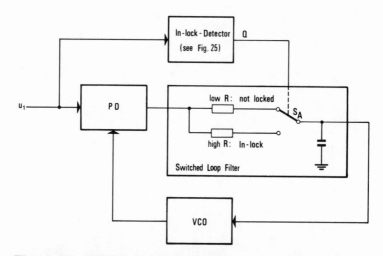

**Fig. 3-23 Simplified block diagram of a PLL using a switched loop filter for the acquisition of noisy signals. ($S_A$ = analog switch.)**

To control the analog switch we need a signal which tells us whether or not the PLL is in the locked state. Such a control signal is generated by the *in-lock detector* shown in Fig. 3-22. The in-lock detector is a cascade connection of a 90° phase shifter, an analog multiplier, a low-pass filter, and a Schmitt trigger. If the PLL is locked, there is a phase offset of approximately 90° between input signal $u_1$ and VCO output signal $u_2$. The phase shifter then outputs a signal $u_1$, which is nearly in phase with the VCO output signal $u_2$. The average value of the output signal of the multiplier $u_M = u_1' \cdot u_2$ is positive. If the PLL is not in the locked state, however, the signals $u_1'$ and $u_2$ are uncorrelated, and the average value of $u_M$ is zero. Thus the output signal of the multiplier is a clear indication of lock. To eliminate ac components and to inhibit false triggering, the $u_M$ signal is conditioned by a low-pass filter. The filtered signal is applied to the input of the Schmitt trigger.

If the PLL has locked, the integrator is kept in the HOLD mode, as mentioned. Of course, an analog integrator would drift away after some time. To avoid this effect, an antidrift circuit has to be added; this is not shown however in Fig. 3-22.

**b. The Switched-Filter Technique**  This locking method is depicted in Fig. 3-23. This configuration uses a loop filter whose bandwidth can be switched by a binary signal. The control signal for the switched filter is also derived from an in-lock detector, as shown in the previous example. In the unlocked state of the PLL, the output signal $Q$ of the in-lock detector is

*Table 3-1. Summary of Parameters and Formulas for Linear PLLs*

| Parameter category | Symbol | Parameter | Definition | Formulas for second-order PLLs |
|---|---|---|---|---|
| General | $\omega_0$ | Center frequency of the VCO | Angular frequency of the VCO at $u_f = 0$ | |
| | $\tau_1, \tau_2$ | Time constants of loop filter | | |
| | $\omega_n$ | Natural frequency of the PLL | $\omega_n$ is the natural frequency of the PLL system. The PLL responds to an excitation at its input with a transient, normally a damped oscillation with the angular frequency $\omega_n$. | |
| | $\zeta$ | Damping factor | $1/\omega_n\zeta$ = time constant of the damped oscillation. | |
| Acquisition | $\Delta\omega_H$ | Hold range | Frequency range within which PLL operation can be statically stable. | Passive filter: $\Delta\omega_H = K_0 K_d$ <br> Active filter: $\Delta\omega_H = \infty$ (theoretically) |
| | $\Delta\omega_L$ | Lock range | If the frequency offset of the reference signal is smaller than the lock range, the PLL locks within one single beat note between reference and output frequencies. | Filter types 1 and 4: $\Delta\omega_L = \omega_n$ <br> Filter types 2 and 3: $\Delta\omega_L = 2\zeta\omega_n$ |
| | $T_L$ | Lock-in time | Time required for the lock-in process. | $T_L \approx \dfrac{1}{\omega_n}$ |

| | Symbol | Name | Description | Formula |
|---|---|---|---|---|
| | $\Delta\omega_P$ | Pull-in range | If the frequency offset of the reference signal is larger than the lock range but smaller than the pull-in range, the PLL will slowly lock after a number of beat notes between reference and output frequencies. | General: $\Delta\omega_P \approx \dfrac{8}{\tau} \times \sqrt{\zeta\omega_n K_0 K_d} - \omega_n^2$  For HG-loops: $\Delta\omega_P \approx \dfrac{8}{\tau} \times \sqrt{\zeta\omega_n K_0 K_d}$ |
| | $T_P$ | Pull-in time | Time required for a pull-in process. | $T_P \approx \dfrac{\Delta\omega_0^2}{2\zeta\omega_n^3}$ |
| Tracking | $\Delta\omega_{PO}$ | Pull-out range | Dynamic limit of stable operation of the PLL. The system unlocks if a frequency step larger than $\Delta\omega_{PO}$ is applied to the reference input. | $\Delta\omega_{PO} \approx 1.8\omega_n(\zeta + 1)$ |
| | $\dot\omega$ | Rate of change of frequency offset | Maximum allowable rate of change of (angular) reference frequency. | $\dot\omega < \omega_n^2$ (see text) |
| Noise | $P_s,\ P_n$ | Signal, noise power | Power of input signal and noise signal applied to the input of a PLL. | |
| | $B_i$ | Prefilter bandwidth | Bandwidth of the prefilter (or the input signal source). | |
| | $B_L$ | Noise bandwidth | Noise bandwidth | $B_L \approx \omega_n\left(\zeta + \dfrac{1}{4\zeta}\right)$ |
| | $(SNR)_i$ | Signal-to-noise ratio of the input signal | | $SNR_i = \dfrac{P_s}{P_n}$ |
| | $(SNR)_L$ | Signal-to-noise ratio of the loop | | $SNR_L = SNR_i \dfrac{B_i}{2B_L}$ |

zero. In this state the bandwidth of the loop filter is so large that the lock range exceeds the frequency range within which the input signal is expected. The noise bandwidth is then too large to enable stable operation of the loop. There is nevertheless a high probability that the PLL will lock spontaneously at some time. To avoid repeated unlocking of the loop, the filter bandwidth has to be reduced instantaneously to a value where the noise bandwidth $B_L$ is small enough to provide stable operation. This is done by switching the loop filter to its low-bandwidth position by means of the $Q$ signal.

### 3-5 SUMMARY OF CHAPTER 3

In analyzing the dynamic performance of the linear PLL we defined a number of key parameters and developed some formulas for them. Since the expressions differ with the type of loop filter used, these results are summarized in Table 3-1.

# 4 THEORY OF THE DIGITAL PLL

We have shown that the linear PLL is constructed entirely from analog functional blocks. For example, an analog multiplier was used for the phase detector.

The *digital PLL* is more difficult to define, because the word digital is applied to a number of different things. In the definition of the classical digital PLL (DPLL), the term digital simply meant that both the reference signal $u_1$ and the output signal $u_2$ (as defined in Fig. 1-1a) are *binary* signals, that is, square-wave signals. Within the classical digital PLL, analog functional blocks are still used, such as $RC$ loop filters, VCOs, and so on. We could state that the classical DPLL is characterized by the appearance of *intermediate analog signals*.

In the last several years DPLL systems have been developed which are built exclusively from digital devices. No intermediate analog signal can be found in these systems. The loop filter, for example, can now no longer be an $RC$ filter, but could be a first- or second-order digital filter. This newer type of DPLL system will be distinguished from the classical one by calling it an *all-digital PLL* (ADPLL) in the following text. DPLLs that have been implemented by software on a microcomputer also belong to this class. It is simpler to treat the classical DPLL and the ADPLL separately. The theory of the DPLL with intermediate analog signals will be discussed in Sec. 4-1, and Sec. 4-2 will be devoted to the all-digital and software-based DPLLs.

## 4-1 DIGITAL PLLs WITH INTERMEDIATE ANALOG SIGNALS

This type of PLL differs from the linear PLL only in the type of PD used. In the case of the linear PLL, the type 1 PD (Table 2-1) has been applied almost exclusively. With most of the classical DPLLs, type 2, 3, or 4 PDs (defined in Table 2-1) are used. Type 4, also referred to as the phase/frequency detector, is the most popular because it permits arbitrarily large lock ranges. As will be demonstrated in Secs. 4-1-2 and 4-1-3, a very simple mathematical model of the DPLL is obtained when the type 4 PD is chosen.

The type of PD used has a strong influence on the dynamic response of the DPLL. Section 4-1-1 will therefore explain how the various types of digital PDs act on the digital PLL.

### 4-1-1 Influence of the Phase-Dectector Type on the Dynamic Performance of the Digital PLL

As shown in Table 2-1, three types of digital phase-detector circuits are generally used:

1. The EXCLUSIVE-OR gate (XOR), type 2

2. The edge-triggered $JK$ flip-flop, type 3

3. The so-called phase/frequency detector, type 4

Many other circuits can be designed, of course, but we will restrict our discussion to these three types.

Let us start with the simplest one, the XOR gate (Fig. 4-1). The circuit is very simple and inexpensive but has a major weakness—its output signal is strongly dependent on the duty-cycle ratios of signals $u_1(t)$ and $u_2(t)$.[9] Let us assume for the moment that these signals have a symmetrical waveform. As was the case with the type 1 PD, the phase difference between reference and output signals is $\pi/2$ (90°) if the PLL is working at its center frequency $\omega_0$ (refer to Fig. 4-1a). The phase error is defined as 0 in this situation, in analogy with the linear PLL. The output signal $u_d(t)$ of the PD is a symmetrical square wave having twice the frequency of the reference signal. The average signal $\overline{u_d}(t)$ is then defined as zero; this corresponds to introducing an offset of half the supply voltage of the XOR gate.[*]

If the phase difference between the $u_1$ and $u_2$ signals becomes smaller, the duty-cycle ratio of the $u_d$ signal will be less than 50 percent, which means that $\overline{u_d}(t)$ becomes negative. Figure 4-1b shows the extreme case where $u_1$ and $u_2$ are in phase and $\overline{u_d}(t)$ is at its lowest. The phase error here by definition is $-\pi/2$.

The other extreme is shown in Fig. 4-1c. Here, the phase error is defined as $+\pi/2$, and $\overline{u_d}(t)$ shows its largest value.

These examples demonstrate that a PLL using an XOR-type PD can, like a linear PLL, operate with a static phase error $\theta_e$ in the range of $-\pi/2 \le \theta_e \le +\pi/2$.

If the average output of the PD is plotted against the phase error $\theta_e$,

---

[*]If the XOR gate is run from a supply voltage of $+5$ V (CMOS or TTL level), its average output level is 2.5 V, which is defined as $\overline{u_d}(t) = 0$.

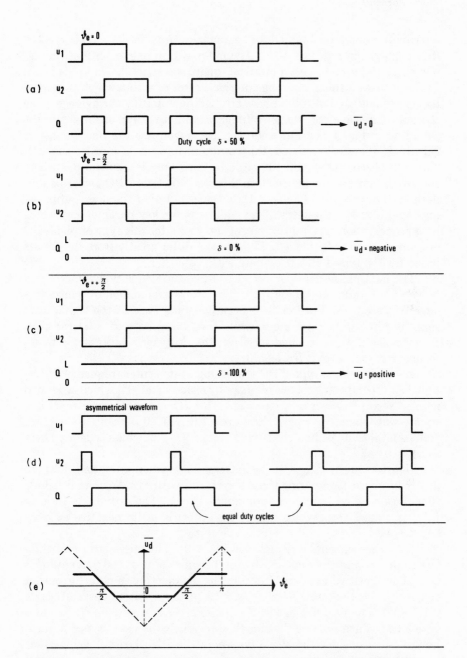

Fig. 4-1  Performance of the type 2 PD (according to Table 2-1). Parts (a)–(e) show the PD's response to various phase errors and duty-cycle ratios of the $u_1$ and $u_2$ signals.

a triangular curve is obtained for symmetrical waveforms $u_1(t)$ and $u_2(t)$. (See dashed curve in Fig. 4-1$e$.) Assume now that one or both of the signals $u_1(t)$ and $u_2(t)$ are asymmetrical square waves, as shown in Fig. 4-1$d$. In this situation there is a range of phase error $\theta_e$ within which the signal $u_d(t)$ does not vary with the phase error. When plotting the average value $\overline{u_d}$ against phase error, we get a function which is clipped. (Refer to the solid line in Fig. 4-1$e$.) For asymmetrical waveforms the phase-detector gain $K_d$ is therefore reduced; this results in lower hold and lock ranges. The XOR-based PD should therefore be used only in applications where the waveforms are symmetrical. In this case, the type 2 PD performs similarly to the type 1 PD (Table 2-1), but does not offer the noise-suppressing capabilities of the latter. The equations for key parameters such as hold range, lock range, pull-in range, and so on, for this type of phase PD are summarized in Table 4-1. They do not differ greatly from the results found for the type 1 PD. (Compare with Table 3-1.)

The next phase-detector circuit to be considered is type 3 (according to Table 2-1; refer also to Fig. 4-2).[9] This PD is exemplified by a negative-edge triggered $JK$ flip-flop. The negative-going edge of the signal $u_1(t)$ sets the flip-flop to its 1 state, while the falling edge of $u_2(t)$ resets it. Because the flip-flop is edge-sensitive, the duty-cycle ratio of the waveforms $u_1(t)$ and $u_2(t)$ is *irrelevant*.

If a PLL using a type 3 PD is working at its center frequency $\omega_0$, the signals $u_1(t)$ and $u_2(t)$ are out of phase by exactly $\pi$ (180°) which is defined here as zero phase error $\theta_e$ (see Fig. 4-2$b$). The output signal $u_d(t)$ is a square-wave signal having a duty-cycle ratio of 50 percent. Its average voltage level is about half the supply voltage ($U_B/2$), which is defined here as $\overline{u_d} = 0$.

Figure 4-2$a$ shows one of the extreme cases where $u_1(t)$ and $u_2(t)$ are nearly in phase. Comparing these waveforms with Fig. 4-2$b$, we conclude that the phase error now is approximately $-\pi$. The duty-cycle ratio of $\overline{u_d}(t)$ is almost zero, which means that $\overline{u_d}(t)$ is at its negative extreme ($\overline{u_d} \approx 0$ V).

The other extreme is depicted in Fig. 4-2$c$. The phase error is slightly less than $+\pi$ and the duty-cycle ratio becomes almost 100 percent. $\overline{u_d}$ is now at its positive extreme. Plotting the average phase-detector output signal $\overline{u_d}(t)$ against phase error $\theta_e$ results in the sawtooth curve of Table 2-1, line 3. The useful range of the phase error is $-\pi < \theta_e < +\pi$ for a type 3 PD, which is twice that for the previously discussed types 1 and 2. Let us now consider the situation where the PLL has not yet become locked, i.e. $\omega_1 \neq \omega_2$.

The example in Fig. 4-2$d$ illustrates the situation where $\omega_1 \gg \omega_2$; the reference frequency is much higher than the frequency of the VCO. The flip-flop is consequently much more frequently set than reset; i.e., it is in the 1 state most of the time. The greater the value of $\omega_1$, the more nearly the duty-cycle ratio of $u_d(t)$ approaches 100 percent. This signifies that the frequency of the VCO will be pulled in the positive direction, which is most desirable.

If on the other hand, $\omega_1 \ll \omega_2$, the reverse becomes true (Figure 4-2e). The duty cycle now approaches zero percent, and the frequency of the VCO is again pulled in the desired direction. Hence the type 3 PD proves to be phase- and frequency-sensitive, which is desirable, but unfortunately the "duality" is not perfect, as will be demonstrated by the following two examples.

Figure 4-2f illustrates the case where the two frequencies $\omega_1$ and $\omega_2$ are close together, that is, $\omega_1 \approx 1.1\omega_2$. The duty-cycle ratio of the signal $u_d(t)$ is then continuously swept from 0 to 100 percent, and the average signal $\overline{u_d}(t)$ shows the form of a sawtooth having a frequency of $\omega_1 - \omega_2$. For $\omega_1 \approx 0.9\omega_2$ (Fig. 4-2g) the situation is similar. The duty-cycle ratio of $u_d(t)$ is swept in the opposite direction, and the average signal $\overline{u_d}(t)$ is again a sawtooth having a frequency $\omega_2 - \omega_1$. The average value $\overline{u_d}(t)$ is near zero in both cases, which indicates that the frequency-sensitive behavior is almost completely lost if the two frequencies $\omega_1$ and $\omega_2$ come closer. If $\overline{u_d}(t)$ is plotted against the frequency offset $\omega_1 - \omega_2$ (see Table 2-1), we obtain a curve that rises monotonically with $(\omega_1 - \omega_2)$, but the function is undefined for small offsets. This is shown by the shaded area in Table 2-1. For $\omega_1 \approx \omega_2$ the type 3 PD behaves very much like the type 1 PD where a pull-in effect onto the VCO was observed because of the unharmonicity of the frequency modulation of the VCO. Table 4-1 also lists the key parameters, such as hold range, lock range, and so on, for the type 3 PD.

The type 4 PD outperforms its two predecessors in every regard because it is also independent of the duty-cycle ratio of the waveforms $u_1(t)$ and $u_2(t)$, and furthermore it exhibits a marked sensitivity to frequency for even the smallest offset $\omega_1 - \omega_2$.[8] This circuit is therefore often referred to as a phase/frequency detector. A typical circuit for this type of PD is shown in Table 2-1, line 4.

As seen from this schematic diagram, the device is basically built from two $RS$ flip-flops and two additional latches which have been designated UP and DOWN, respectively. We must remember that both latches have active-low outputs. Moreover the latches are set (or reset) by the falling edges of $u_1(t)$ and $u_2(t)$, respectively. The operation of the type 4 PD is as follows:

1. If the rising edge of $u_1(t)$ occurs at a time when $u_2(t)$ is low, the next falling edge of $u_1(t)$ will set the UP latch LOW (1 state), and the next falling edge of $u_2(t)$ will reset this latch HIGH (0 state).

### Table 4-1. Summary of the most important design equations for the

| Parameter | | Type 2 PD | |
|---|---|---|---|
| | | With active loop filter | With passive loop filter |
| Hold range, $\Delta\omega_H$ | | $\Delta\omega_H \to \infty$ | $\Delta\omega_H = \dfrac{\pi}{2}\dfrac{K_0 K_d}{N}$ |
| Lock range, $\Delta\omega_L$ | Filter having a zero $\tau_2 \neq 0$ | $\Delta\omega_L \approx \pi\zeta\omega_n$ | |
| | Filter without a zero $\tau_2 = 0$ | $\Delta\omega_L \approx \dfrac{\pi}{\sqrt{8}}\,\omega_n$ | |
| Pull-in range, $\Delta\omega_P$ | | $\Delta\omega_P \approx \dfrac{\pi}{2}\sqrt{\dfrac{2\zeta\omega_n K_0 K_d}{N}}$ | $\Delta\omega_P \approx \dfrac{\pi}{2}\sqrt{\dfrac{2\zeta\omega_n K_0 K_d}{N} - \omega_n^2}$ |
| Pull-in time, $T_P$ | | $T_P \approx \dfrac{4}{\pi^2}\dfrac{\Delta\omega_0^2}{\zeta\omega_n^3}$ | |
| Pull-out range, $\Delta\omega_{PO}$ | $\zeta < 1$ | $\Delta\omega_{PO} \approx 1.8\omega_n(\zeta + 1)$ | |
| | $\zeta > 1$ | | |

**nput** of the PLL.

| Type 3 PD | | Type 4 PD |
|---|---|---|
| **With active loop filter** | **With passive loop filter** | |
| $\Delta\omega_H \to \infty$ | $\Delta\omega_H = \pi \dfrac{K_0 K_d}{N}$ | $\Delta\omega_H \to \infty$ |
| $\Delta\omega_L \approx 2\pi\zeta\omega_n$ | | Not defined |
| $\Delta\omega_L \approx \dfrac{\pi}{\sqrt{8}}\,\omega_n$ | | |
| $\omega_P \approx \pi \ \sqrt{\dfrac{2\zeta\omega_n K_0 K_d}{N}}$ | $\Delta\omega_P \approx \pi \ \sqrt{\dfrac{2\zeta\omega_n K_0 K_d}{N} - \omega_n^2}$ | $\Delta\omega_P \to \infty$ |
| $T_P \approx \dfrac{\Delta\omega_0^2}{\pi^2\zeta\omega_n^2}$ | | $T_P \approx \dfrac{N\tau_1 + \tau_2}{K_0 K_d'}$ |
| $\Delta\omega_{PO} = \pi\omega_n \exp\left[ \dfrac{\zeta}{\sqrt{1-\zeta^2}} \tan^{-1}\dfrac{\sqrt{1-\zeta^2}}{\zeta} \right]$ | | $\Delta\omega_{PO} = 2\pi\omega_n \exp\left[ \dfrac{\zeta}{\sqrt{1-\zeta^2}} \tan^{-1}\dfrac{\sqrt{1-\zeta^2}}{\zeta} \right]$ |
| $\Delta\omega_{PO} = \pi\omega_n \exp\left[ \dfrac{\zeta}{\sqrt{\zeta^2-1}} \tanh^{-1}\dfrac{\sqrt{\zeta^2-1}}{\zeta} \right]$ | | $\Delta\omega_{PO} = 2\pi\omega_n \exp\left[ \dfrac{\zeta}{\sqrt{\zeta^2-1}} \tanh^{-1}\dfrac{\sqrt{\zeta^2-1}}{\zeta} \right]$ |

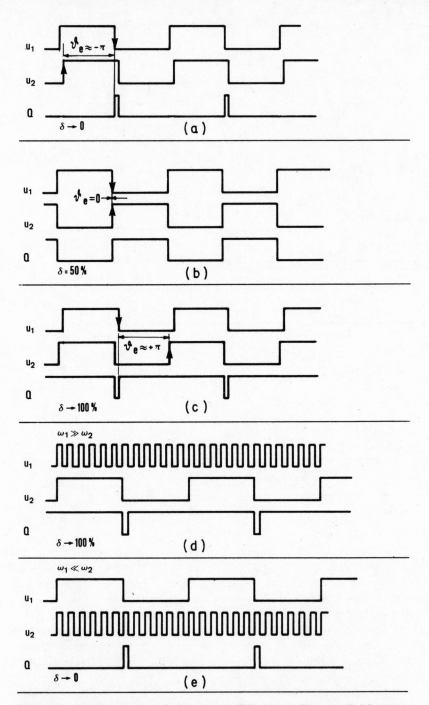

**Fig. 4-2** Performance of the type 3 PD (according to Table 2-1). Parts (*a*)-(*g*) are seven different cases of phase and frequency offset.

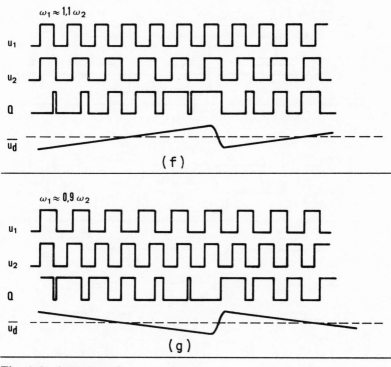

**Fig. 4-2**   (*continued*)

2. If the rising edge of $u_1(t)$ occurs at a time when $u_2(t)$ is high, then the next falling edge of $u_1(t)$ will set the DOWN latch LOW (1 state), and the next falling edge of $u_2(t)$ will reset this latch HIGH (0 state).

Figure 4-3 explains the phase-and frequency-sensitive performance of the type 4 PD in more detail. Four different cases are considered; for the first two it is assumed that $\omega_1 = \omega_2$. If $u_2(t)$ has the same frequency as $u_1(t)$ but is lagging behind $u_1(t)$ (Fig. 4-3a), the output of the UP latch is pulsed in proportion to the phase error $\theta_e$. The DOWN latch is in its inactive state. This causes the frequency of the VCO to be momentarily increased, so that the phase error is reduced to zero. In normal operation there is always some small phase jitter between $u_1(t)$ and $u_2(t)$, and the frequency of the VCO is continuously corrected up and down.

Figure 4-3b shows the case where $u_2(t)$ is leading $u_1(t)$. Now the input of the DOWN latch is pulsed in proportion to the phase error $\theta_e$, and the UP latch remains in its inactive state here. This causes the frequency of the VCO to be decreased temporarily until the phase error is reduced to near zero.

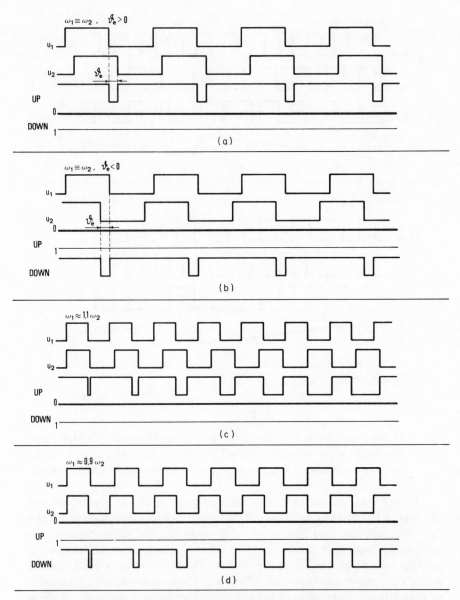

**Fig. 4-3** Performance of the type 4 PD (according to Table 2-1). Parts (a)–(d) are four different cases of phase and frequency offset.

The next two cases illustrate frequency-sensitive performance. Figure 4-3c shows the situation where $\omega_1 \approx 1.1\omega_2$. It is clearly seen that now only the UP latch is pulsed, the DOWN latch staying inactive all the time. Only the DOWN latch is pulsed, however, if $\omega_1$ is smaller than $\omega_2$, as shown for $\omega_1 \approx 0.9\omega_2$ in Fig. 4-3d. If the UP and DOWN outputs of the PD are used to drive a charge-integrating circuit (usually called a charge pump; see also Table 2-1, col. 7),[10] the frequency of the VCO will always be pulled in the correct direction. The pull-in range and the hold range of a PLL that makes use of such phase/frequency detector must therefore, at least theoretically, be infinite. Hence, it is not astonishing that this type of PD is preferred in most digital PLL applications. The average output signal $\overline{u_d}(t)$ of the type 4 PD is defined as the weighted duty-cycle ratio of the UP and DOWN outputs; a weight of $+1$ is assigned to the output of the UP latch and a weight of $-1$ to that of the DOWN latch.

### 4-1-2 The Lock-In Process of the PLL

Figure 4-4a shows a generalized block diagram of a PLL using a type 4 PD. Besides the three already known functional blocks, a frequency divider has been added. PLLs are used very often in frequency synthesis. The PLL then has to generate an output signal whose frequency is $N$ times that of the reference frequency, where $N$ is the scaling factor of the frequency divider. If $N$ is set at 1, the well-known configuration of Fig. 1-1a is again obtained.

In the circuit of Fig. 4-4a the PD no longer directly compares the reference signal $u_1(t)$ with the output signal $u_2(t)$, but compares it with the scaled-down output signal instead. In the locked state the reference and output frequencies are thus no longer identical, but differ by a factor of $N$. If we speak of the hold, lock, or pull-out range of such a system, we first have to decide whether these parameters are related to the reference input of the PLL or to its output. In the following discussion we will always specify these parameters in relation to the reference input.

In analogy to the linear PLL, formulas for the parameters such as hold range $\Delta\omega_H$, lock range $\Delta\omega_L$, pull-in range $\Delta\omega_P$, pull-out range $\Delta\omega_{PO}$, pull-in time $T_P$, and so on, can be derived. A detailed mathematical treatment would be too lengthy, hence only the final results are listed in Table 4-1. If these expressions are compared with those obtained for the linear PLL (Table 3-1), we conclude that many formulas are quite similar, but differ in the proportionality factors. In many texts and application notes this is completely disregarded; many authors furthermore are using formulas obtained for the linear PLL to design any kind of digital PLL, not being aware that there might be fundamental differences.

Indeed the formulas for digital PLLs with type 2 or 3 PDs are not too

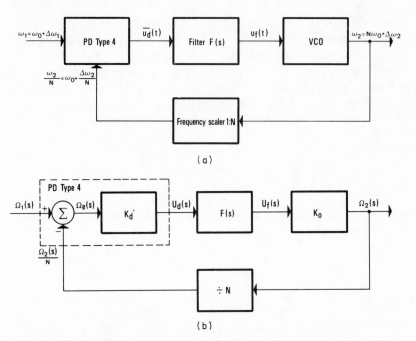

(a)

(b)

**Fig. 4-4  If a DPLL uses a type 4 PD, a linear model can be derived even for the unlocked state of the system. (*a*) Block diagram of the system in the time domain. (*b*) Mathematical model of the system in the complex-frequency domain.**

different from those valid for the linear PLL, but the digital PLL utilizing a type 4 phase/frequency detector shows a completely different dynamic performance. This stems from the unique properties of the phase-detector circuit, as will be demonstrated by the following considerations. Returning to the block diagram of the DPLL in Fig. 4-4*a*, we will see immediately that a linear mathematical model can be developed even for the *unlocked* state when a type 4 PD is used. Let us first define the following variables:

$$\omega_1 = \omega_0 + \Delta\omega_1 = \text{reference frequency} \tag{4-1a}$$

where $\omega_0$ is the center frequency referred to the *input*,

$$\omega_2 = N\omega_0 + \Delta\omega_2 = \text{output frequency} \tag{4-1b}$$

where $N\omega_0$ is the center frequency referred to the *output*, and

$$\omega_e = \omega_1 - \omega_2/N = \text{(angular) frequency error} \tag{4-1c}$$

which is referred to the *input*.

Furthermore we will define the Laplace transforms of these variables as follows:

$$\Omega_1(s) = \mathcal{L}\{\Delta\omega_1(t)\} \tag{4-2a}$$

$$\Omega_2(s) = \mathcal{L}\{\Delta\omega_2(t)\} \tag{4-2b}$$

$$\Omega_e(s) = \mathcal{L}\{\omega_e(t)\} \tag{4-3c}$$

In the locked state ($\omega_1 = \omega_2/N$), the average output signal $\overline{u_d}(t)$ of the PD, is dependent upon the phase error $\theta_e$,

$$\overline{u_d} = K_d\theta_e \tag{4-3}$$

According to Table 2-1 this relation is valid in the range $-2\pi < \theta_e < +2\pi$. We remember that $K_d$ has been called detector gain. More precisely, $K_d$ should be referred to as *detector gain in the locked state*.

In the unlocked state $\overline{u_d}(t)$ has been shown to depend on the frequency error $\omega_e$. For a rough approximation, we can say that $\overline{u_d}$ is now proportional to the frequency error $\omega_e$,

$$\overline{u_d} \approx K_d'\omega_e = K_d'\left(\omega_1 - \frac{\omega_2}{N}\right) \tag{4-4}$$

This equation is derived in Appendix D.

The primed coefficient $K_d'$ in Eq. (4-4) is called the phase-dectector gain in the unlocked state and should not be confused with the previously defined coefficient $K_d$. $K_d'$ has the dimension (voltage/angular frequency), i.e., Vs, whereas $K_d$ has the dimension V only.

With Eq. (4-4) and the previously derived Eqs. (3-1) we now have the three equations specifying the dynamic performance of the DPLL. To recall these, we have

- For the PD: $\qquad\qquad U_d(s) = K_d'\Omega_e(s) \tag{4-5a}$

- For the loop filter: $\qquad U_f(s) = U_d(s) \cdot F(s) \tag{4-5b}$

- For the VCO: $\qquad\quad \Omega_2(s) = K_0U_f(s) \tag{4-5c}$

After eliminating $U_d(s)$ and $U_f(s)$ we get

$$\Omega_e(s) = \frac{N}{N + K_0K_d'F(s)}\,\Omega_1(s) \tag{4-6}$$

The corresponding mathematical model of the DPLL is shown in Fig. 4-4b. This model allows us to calculate the error frequency $\omega_e$ as a function of time for any exciting function applied to the input of the PLL. Let us perform the computation for the case where the frequency of the reference signal $\omega_1$ is constant but is offset by the amount $\Delta\omega$ from the center frequency $\omega_0$. It is further assumed that the PLL is unlocked initially and the momentary angular frequency of the VCO is $\omega_0$ at $t = 0$. We consequently, for $\omega_1(t)$ have

$$\omega_1(t) = \omega_0 + \Delta\omega_1 \tag{4-7}$$

For its Laplace transform we get

$$\Omega_1(s) = \frac{\Delta\omega_1}{s} \tag{4-8}$$

For the loop filter an active filter (type 3 in Table 2-2) is chosen. As already mentioned in Chap. 2, this filter type—more frequently referred to as the charge pump—is preferred in conjunction with the type 4 PD (Table 2-1), because it conveniently integrates the charge pulses (of both polarities) delivered by this circuit. Consequently, we obtain

$$F(s) = \frac{1 + s\tau_2}{s\tau_1} \tag{4-9}$$

Combining Eqs. (4-6), (4-8), and (4-9) yields

$$\Omega_e(s) = \frac{N\tau_1/K_0K_d'}{1 + s[\tau_2 + N\tau_1/K_0K_d']} \frac{\Delta\omega_1}{s} \tag{4-10}$$

The response $\omega_e(t)$ of the DPLL is obtained by transforming Eq. (4-10) back into the time domain,

$$\omega_e(t) = \frac{\Delta\omega_1 N\tau_1}{K_0K_d'} e^{-t/T_P} \tag{4-11}$$

Comparing Eqs. (4-10) and (4-11), we obtain for the time constant of the pull-in process

$$T_P = \tau_2 + \frac{N\tau_1}{K_0K_d'} \tag{4-12}$$

Equation (4-11) tells us that the digital PLL using a type 4 PD always gets locked for any frequency offset $\Delta\omega_1$ of the reference signal.

For the hold range and the lock range we get, at least theoretically,

$$\Delta\omega_H = \infty \tag{4-13}$$

$$\Delta\omega_P = \infty \tag{4-14}$$

This was to be expected from the discussion of the phase/frequency detector in Sec. 4-1-1 (see also Fig. 4-3).

What about the lock range of this type of DPLL? In the case of the *linear* PLL, the lock range $\Delta\omega_L$ has been the maximum frequency offset allowing the PLL to get locked within one single-beat note between reference and output signals. If the linear PLL was not yet locked, the frequency of the VCO was modulated in the positive and negative directions around its center frequency $\omega_0$. If the peak frequency deviation of the VCO was large enough, the system could become locked rapidly.

The situation is very different in the case of the *digital* PLL. As the preceding analysis has shown, the average output signal of the phase/frequency detector $\overline{u_d}$ is not an ac signal if $\omega_1 \neq \omega_2/N$, but *always* has a positive polarity for $\omega_1 > \omega_2/N$ or a negative polarity for $\omega_1 < \omega_2/N$. Hence the frequency of the VCO is never modulated around its center frequency, but is always pulled in *one single direction,* namely, in the direction of the reference frequency. Strictly speaking, it is not even possible to define a lock range for this type of DPLL; the lock-in process is essentially always a pull-in process.

Designing a digital PLL is not much different from designing a linear one, because the key parameters, such as hold range or pull-out range, are easily computed by using the formulas given in Table 4-1. If we choose type 2 or 3 PD, we do not encounter any problem, since the numerical data for parameters such as $K_0$ and $K_d$ are found on every data sheet. The parameter $K_d'$ is not found on data sheets, because $K_d'$ is a *new* parameter, which is defined for the first time in this book. The performance of the type 4 PD will be analyzed in detail in Appendix D. A simple equation for $K_d'$ will be derived there:

$$K_d' \approx \frac{U_B}{\omega_0} \tag{4-15}$$

In this equation $U_B$ is the supply voltage of the type 4 PD, and $\omega_0$ is the center frequency of the PLL related to its input.

The preceding discussion has shown that the lock-in process of the digital PLL is much easier to understand than the corresponding process for the linear PLL. For the designer of practical circuits it is sufficient to know that the hold and pull-in ranges are as large as the operating frequency range of the PLL, and that the time constant of the pull-in process is simply given by Eq. (4-12).

### 4-1-3 Tracking Performance of the DPLL

Now that the lock-in process of DPLLs in Sec. 4-1-2 has been analyzed, the tracking performance remains to be considered. As with the linear PLL (Chap. 3), this can be done by calculating the transient response of the phase signal $\theta_2(t)$ or of the phase error $\theta_e(t)$. In the locked state we

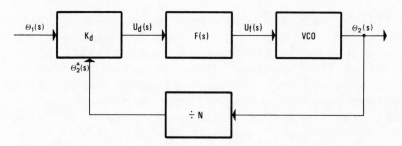

**Fig. 4-5 Linear mathematical model of the DPLL for the locked state.**

had $\omega_1 = \omega_2/N$ (Fig. 4-4a). We can draw the mathematical model of Fig. 4-5 for the phase signals. This model is almost identical to that of the linear PLL (Fig. 3-2), except that there is an additional frequency divider. If $\theta_2(t)$ is defined as the phase signal at the output of the VCO, $\theta_2^*(t) = \theta_2(t)/N$ is the phase signal at the output of the frequency divider.

The Laplace transform of $\theta_2(t)$ is then

$$\Theta_2(s) = \mathcal{L}\{\theta_2(t)\}$$

while the Laplace transform of $\theta_2^*(t)$ is, by analogy,

$$\Theta_2^*(s) = \mathcal{L}\{\theta_2^*(t)\} = \frac{1}{N}\mathcal{L}\{\theta_2(t)\}$$

The phase error $\theta_e(t)$ is now defined as

$$\theta_e(t) = \theta_1(t) - \frac{\theta_2(t)}{N} \tag{4-16a}$$

whereas its Laplace transform is given by

$$\Theta_e(s) = \Theta_1(s) - \frac{\Theta_2(s)}{N} \tag{4-16b}$$

The phase-transfer function $H(s)$ of the digital PLL is defined by

$$H^*(s) = \frac{\Theta_2(s)}{\Theta_1(s)} \tag{4-17}$$

and the error-transfer function is

$$H_e^*(s) = \frac{\Theta_e(s)}{\Theta_1(s)} \tag{4-18}$$

On the basis of the mathematical model of Fig. 4-5 we obtain for $H^*(s)$ and $H_2^*(s)$

$$H^*(s) = \frac{K_0 K_d F(s)}{Ns + K_0 K_d F(s)} \tag{4-19}$$

$$H_e^*(s) = \frac{Ns}{Ns + K_0 K_d F(s)} \tag{4-20}$$

In the case of the DPLL using the type 4 PD, the type 3 loop filter (Table 2-2), usually called a charge pump, is most frequently chosen. For the transfer function of this loop filter (Table 2-2) we already have

$$F(s) = \frac{1 + s\tau_2}{s\tau_1}$$

Substituting this expression into Eqs. (4-19) and (4-20) yields

$$H^*(s) = \frac{(K_0 K_d / N\tau_1)(1 + s\tau_2)}{s^2 + s\,(\tau_2 K_0 K_d / \tau_1 N) + (K_0 K_d / N\tau_1)} \tag{4-21}$$

$$H_e^*(s) = \frac{s^2}{s^2 + s\,(\tau_2 K_0 K_d / \tau_1 N) + (K_0 K_d / N\tau_1)} \tag{4-22}$$

As with the linear PLL, we can bring these two expressions into the normalized form by substituting

$$\omega_n = \sqrt{\frac{K_0 K_d}{N\tau_1}} \tag{4-23a}$$

$$\zeta = \frac{\tau_2}{2} \sqrt{\frac{K_0 K_d}{N\tau_1}} \tag{4-23b}$$

For $H^*(s)$ and $H_e^*(s)$ we then obtain the expressions

$$H^*(s) = \frac{2\zeta\omega_n s + \omega_n^2}{s^2 + 2\zeta\omega_n s + \omega_n^2} \tag{4-24}$$

$$H_e^*(s) = \frac{s^2}{s^2 + 2\zeta\omega_n s + \omega_n^2} \tag{4-25}$$

which prove to be identical with those derived for the linear PLL using an active loop filter [refer to Eqs. (3-8) and (3-9)].

This means that the tracking performance of the digital PLL is the same as that for the linear PLL. The dynamic response of the DPLL to

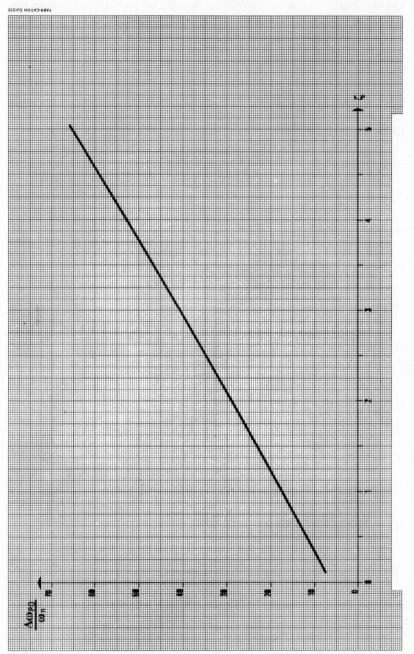

**Fig. 4-6  Pull-out range of the DPLL (type 4 PD) as a function of ζ.**

phase- or frequency-modulated reference signals can therefore be taken from the Figs. 3-14 to 3-16.

As with the linear PLL, we can also try to find expressions for the pull-out range $\Delta\omega_{PO}$. The average output signal $\overline{u_d}(t)$ as a function of phase error $\theta_e$ of the type 2 PD looks very much like that for the type 1 PD; hence we can use the same approximating equation for the pull-out range $\Delta\omega_{PO}$ (Table 4-1).

For the type 3 PD, $\overline{u_d}(t)$ is a linear function of $\theta_e$ in the range $-\pi < \theta_e < +\pi$. The system will get unlocked whenever the phase error exceeds $\pi$. Therefore the pull-out range can be calculated exactly, as is demonstrated in Appendix E. The same holds true for type 4 PD, with the only difference that the peak phase error is $2\pi$ and not $\pi$.

The pull-out range of a DPLL using a type 4 PD is of course dependent on the damping factor $\zeta$. Figure 4-6 shows the pull-out range of this DPLL as a function of $\zeta$.

If we use a PLL for frequency synthesis, we must be aware of the fact that the damping factor $\zeta$ is also a function of the divider ratio $N$, as indicated by Eq. (4-23b). If the divider ratio is variable in the range $N_{min} \leq N \leq N_{max}$, we obtain, for the ratio of maximum-to-minimum damping factor,

$$\frac{\zeta_{max}}{\zeta_{min}} = \sqrt{\frac{N_{max}}{N_{min}}} \qquad (4\text{-}26)$$

If $N$ is changed by a factor of say 100, $\zeta$ varies by a factor of 10, which would be intolerable. The PLL would be underdamped or overdamped at the limits of its frequency range. In such an application the time constants of the loop filter have to be switched in accordance with the instantaneous divider ratio in order to achieve a damping factor in the approximate range between 0.5 and 1.

### 4-2 ALL-DIGITAL AND SOFTWARE-BASED PLLs

As stated in the introduction to Chap. 4, the ADPLL is a system which is built exclusively from digital functional blocks. The term digital is still ambiguous in this context: Digital can mean that an input or output signal of a particular functional block can show up only the values HIGH or LOW, and hence is a *binary* signal; but digital can also signify that a particular signal is an $n$-bit digital word and can represent $2^n$ different states. In Sec. 4-2-1 we will therefore first review the various kinds of block structures encountered in ADPLL systems.

### 4-2-1 Structure of Functional Blocks in ADPLLs

The building blocks of ADPLLs can have very different forms (Fig. 4-7). As can be seen in Fig. 4-7*a*, the input signal (or signals) could be given by one or more *binary* signals; the same could hold true for its output (or outputs). The XOR PD is an example of such a block; it consists of two binary input signals and one binary output signal. The input of a functional block could be an *n*-bit word as well (Fig. 4-7*b*), and the output could still be given by one or more binary signals. (In the following illustrations, single-bit digital signals will be represented by a single line, *n*-bit word signals by a double line.)

As shown in Fig. 4-7*c*, the situation can be reversed, the input being given by binary signals and the output being an *n*-bit word. An example of this is a reversible *n*-bit binary counter: The counting clock and the UP-DOWN signal are the two binary input signals, and the content of the counter is the *n*-bit output word. Finally, both input and output signals can be *n*-bit word signals (Fig. 4-7*d*). An example of this configuration is the well-known digital filter, to be discussed later in this section.

The logic and/or arithmetic operations within the building blocks of Fig. 4-7 can be executed by either hardware or software under the control of a microcomputer. An arithmetic operation, such as a multiplication,

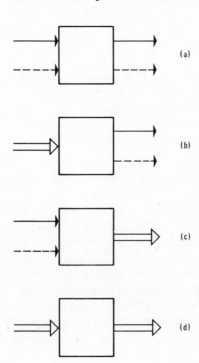

(a)

(b)

(c)

(d)

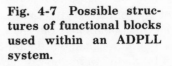

Fig. 4-7 Possible structures of functional blocks used within an ADPLL system.

can be performed by a hardware multiplier; alternatively, the operation can be done by software.

Due to the large impact of microprocessors, software implementations are gaining increased interest. Although the operating speed of microprocessors is improving, it is still the major bottleneck of software-based systems.

A pure hardware digital multiplier ($16 \times 16$ bits), such as a multiplier built as a Schottky transistor-transistor logic integrated circuit (TTL IC) performs a multiplication in a time well below 100 ns. When the same operation is executed on a single-chip microcomputer, a program of some dozens of instructions has to be run. Because typical execution time for one instruction is on the order of 2–3 $\mu$s, the entire operation easily takes 100 $\mu$s or more. (It should be kept in mind that Multiply instructions are normally not available with the simpler microcomputer chips; hence multiplication must be performed by a series of shifted additions.) Limited operating speed is the reason why pure software implementations of PLLs have been successful only for low-frequency applications.[11] The development of faster microprocessors, including fast coprocessors,* will undoubtedly increase the importance of software-based PLLs.

In the Secs. 4-2-2 to 4-2-4 we will review various methods of designing all-digital and software-based functional blocks of PLLs, such as PDs, loop filters, and digtal-controlled oscillators (DCOs). (The DCO is the all-digital counterpart of the classical VCO.) Finally we will discuss some complete ADPLL systems in Sec. 4-2-5.

### 4-2-2 All-Digital PDs

A number of all-digital PDs have already been introduced in Chap. 2. If we consider Table 2-1 again, we easily recognize that the XOR (type 2), the *JK* flip-flop (type 3) and the phase/frequency detector (type 4) are members of this category.

If $n$-bit digital words are admitted for input and/or output signals, a variety of other phase-detector configurations can be devised. The creativity of designers sets practically no limit to the variety of alternatives; we shall therefore restrict our discussion to the most obvious solutions.

A logical evolution of the simple *JK* flip-flop (type 3) PD is the *FF-counter–phase detector* illustrated in Fig. 4-8*a*. The corresponding waveforms are shown in Fig. 4-8*b*.

---

*A coprocessor is an additional chip used in conjunction with microprocessors or single-chip microcomputers. The coprocessor is optimized for the (relatively) fast execution of fixed-and/or floating-point arithmetic operations ($+$, $-$, $\times$, $\div$), eventually including trigonometric functions, log, antilog, and so on.

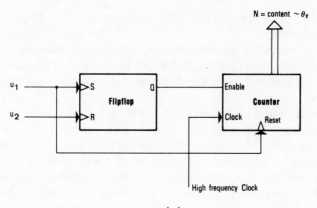

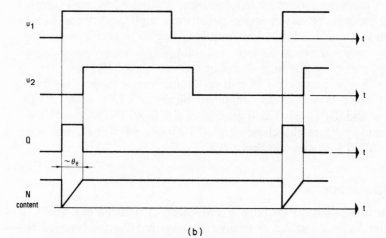

**Fig. 4-8 Flip-flop-counter PD. (*a*) Block diagram. (*b*) Corresponding waveforms.**

The reference (input) signal $u_1$ and the output (or scaled-down output) signal $u_2$ of the DCO (or VCO) are binary-valued signals. They are used to set or reset an edge-triggered *RS* flip-flop. The time period in which the *Q* output of the flip-flop is a logic 1 is proportional to the phase error $\theta_e$. The *Q* signal is used to gate the high-frequency clock signal into the (upward) counter. Note that the counter is reset on every positive edge of the $u_1$ signal.

The content *N* of the counter is also proportional to the phase error $\theta_e$, where *N* is the *n*-bit output of this type of phase detector. The frequency of the high-frequency clock is usually $2^M f_0$, where $f_0$ is the frequency of the reference signal and *M* is a positive integer. Then $2^M$ is the number of quantization levels for the phase error over the range of $0 \leq \theta_e \leq 2\pi$.

Another all-digital phase-detector circuit has become known by the name Nyquist-rate phase detector (NRPD).[12,13] The name stems from the well-known Nyquist theorem which states that a sampled signal can be reconstructed only when the sampling rate is at least twice the highest frequency component of the signal. The block diagram of the NRPD is shown in Fig. 4-9$a$, and the corresponding waveforms are seen in Fig. 4-9$b$.

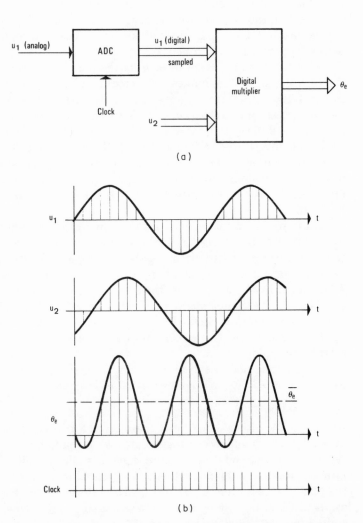

Fig. 4-9   Nyquist-rate PD. ($a$) Block diagram. ($b$) Corresponding waveforms.

The input signal should be an analog signal, such as a signal transmitted over a data line. It is periodically sampled and digitized at the clock rate. In the example shown, the clock rate chosen is 16 times the signal frequency, which is 8 times the Nyquist rate. The signal $u_2$ is an $N$-bit digital word, generated by a DCO. (Refer also to Sec. 7-4.) Furthermore, the signal $u_2$ has been drawn as a sine wave in Fig. 4-9b; another waveform (such as a square wave) could be used as well. The digitized signal $u_1$ and the signal $u_2$ are multiplied together by a software multiplying program. Thus the NRPD operates similarly to the linear (type 1) PD introduced in Chap. 2.

The resulting phase error signal $\theta_e$ is also shown in Fig. 4-9b. Its average value $\overline{\theta_e}$ will have to be filtered out by a succeeding digital loop filter, as will be demonstrated in Sec. 4-2-3.

Still another method of measuring the phase error is the *zero-crossing technique*.[14,15] The simplest zero-crossing phase detector is illustrated in Fig. 4-10a; its waveforms are shown in Fig. 4-10b. The reference signal $u_1$ is supposed to be analog; $u_2$ is a binary signal. The positive transitions of $u_2$ are used to clock the analog-to-digital converter (ADC), so $u_1$ is sampled once during every reference period. The digital output signal of the ADC is then proportional to the phase error. Usually this signal is held in a buffer register until the next conversion is completed; thus the phase-error signal $\theta_e$ is a quasicontinuous signal, as shown by the dashed line in Fig. 4-10b.

An improved version of the zero-crossing phase detector is shown in Fig. 4-11a with the corresponding waveforms in Fig. 4-11b. In contrast to the previously discussed circuit, $u_1$ is sampled here on the positive and negative transitions of $u_2$, doubling the number of phase-error samples. As depicted by the waveforms, the samples taken on the negative edges of $u_2$ have to be complemented (inverted in polarity), which is done by the complementation circuit in Fig. 4-11a. By doubling the number of phase-error samples, this circuit can detect fast variations of the phase error more rapidly.

Another method of extracting the phase error $\theta_e$ is the Hilbert transform.[16] Assume that the reference signal $u_1$ is given by

$$u_1 \cos (\omega_0 + \theta_e)$$

where $\theta_e$ is the hitherto unknown phase error. Furthermore the DCO used in our system is supposed to generate two output signals, an *in-phase* signal $I = \cos \omega_0 t$ and a *quadrature* signal $Q = \sin \omega_0 t$. The Hilbert transform is a relatively simple mathematical procedure of directly calculating the phase error $\theta_e$ from $u_1$, $I$, and $Q$. To perform the Hilbert transform, we require an auxiliary signal $\hat{u}_1(t)$, which is simply the signal $u_1(t)$ phase-shifted by $-90°$ $(-\pi/2)$:

$$\hat{u}_1(t) = u_1 \underline{/-\pi/2} = \cos (\omega_0 t + \theta_e - \pi/2)$$

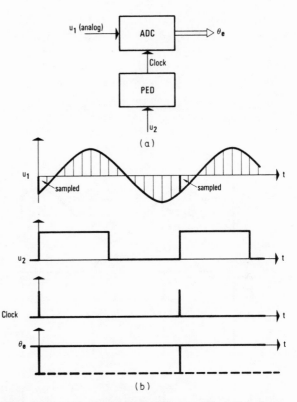

**Fig. 4-10** **Zero-crossing PD sampling the phase error at the positive transitions of the reference signal only. (*a*) Block diagram. (PED = positive-edge detector.) (*b*) Corresponding waveforms.**

We will immediately see how the phase shift is realized. It can be shown, by simple trigonometric manipulations, that

$$\cos \theta_e = Iu_1 + Q\hat{u}_1$$

$$\sin \theta_e = I\hat{u}_1 - Qu_1$$

Figure 4-12*a* shows how these mathematical operations are performed. Note that all signals within the system are supposed to be $N$-bit digital words.

Using a divider, we obtain the tangent of $\theta_e$. Performing the inverse tangent function $\tan^{-1}$ yields the phase error $\theta_e$. The waveforms of the Hilbert-transform phase detector are shown in Fig. 4-12*b*. As in the NRPD, all mathematical computations are performed under the control of a clock signal whose frequency is usually $2^M$ the signal frequency $f_0 = \omega_0/2\pi$. The required phase shift of $-\pi/2$ is then easily obtained by appro-

93

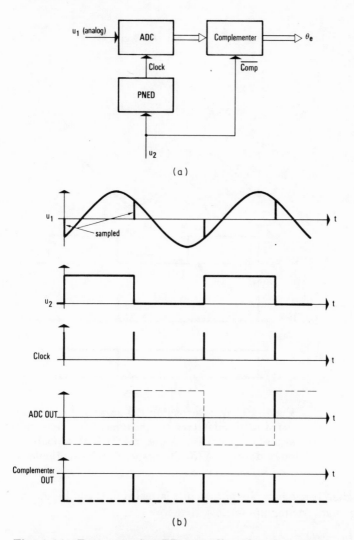

(a)

(b)

**Fig. 4-11   Zero-crossing PD sampling the phase error at both transitions of the reference signal. (*a*) Block diagram. (PNED = positive- and negative-edge detector.) (*b*) Corresponding waveforms.**

priately delaying the signal $u_1(t)$ by a number of shift registers. In the example of Fig. 4-12*b* the clock rate chosen is 16 times the signal frequency, hence the $-\pi/2$ shift would be identical to a digital delay by four clock periods.

The variety of mathematical operations required by the Hilbert transform strongly suggests implementation of this method by software.

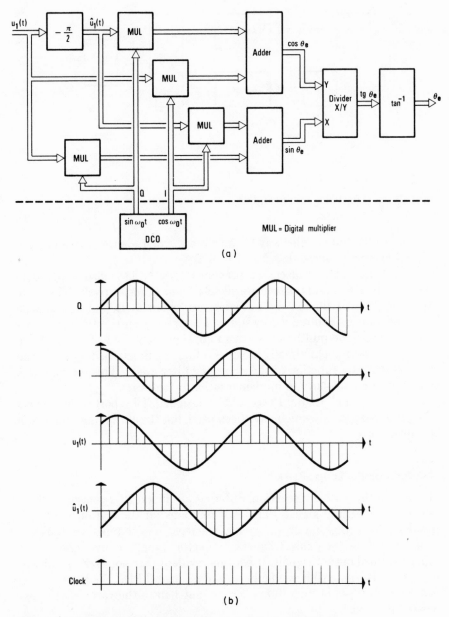

Fig. 4-12 Hilbert-transform PD. (a) Block diagram.
(b) Corresponding waveforms.

95

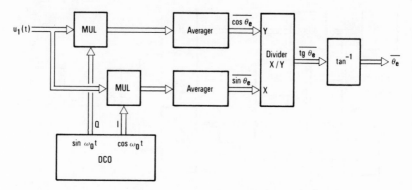

**Fig. 4-13  Digital-averaging PD.**

A similar, but simpler way to calculate the phase error is given by the digital-averaging phase detector (Fig. 4-13).

As in the method discussed previously, the DCO is also required to generate in-phase and quadrature signals $I$ and $Q$, respectively. These are again multiplied by the digital reference signal $u_1(t)$, but the signals $\cos \theta_e$ and $\sin \theta_e$ are obtained by simply averaging (or integrating) the output signals of the multipliers over an appropriate period of time.[12] Note that this arrangement already includes a filtering function, defined by the impulse transfer function of the averaging filter used.[17] This method, too, lends itself particularly to implementation by software.

Many other all-digital PDs could be designed. The short review covers the principal systems which have been used, but the author does not claim completeness.

### 4-2-3 All-Digital Loop Filters

As seen in the preceding section, different all-digital PDs generate different types of output signals. The PDs discussed at the end of Sec. 4-2-2 produce $N$-bit digital output signals, whereas simpler types, such as the XOR (type 2) or the phase/frequency detector (type 4) deliver one or two binary-valued output signals. It becomes evident that not every all-digital loop filter is compatible with all types of all-digital PDs. We have to consider which types of loop filters can be matched to the various PDs discussed previously.

Some ADPLL systems use no loop filter at all. This usually happens in first-order digital PLL systems. In most cases, however, this is not desirable, since the slightest phase error immediately affects the DCO. Loop filters are therefore employed in most cases.

Probably the simplest loop filter is built from an ordinary UP/DOWN counter (Fig. 4-14a). The UP/DOWN counter loop filter preferably operates

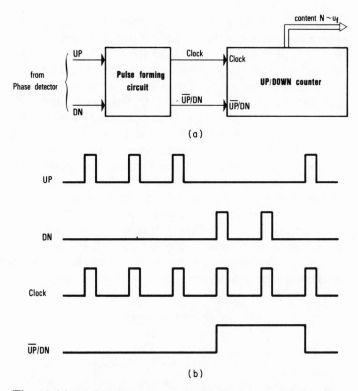

(a)

(b)

**Fig. 4-14** UP/DOWN-**counter loop filter.** (*a*) **Block diagram.** (*b*) **Corresponding waveforms.**

in combination with a phase detector delivering UP or DN (DOWN) pulses, such as the type 4 phase detector in (Table 2-1). It is easily adapted, however, to operate in conjunction with the XOR or *JK* flip-flop phase detectors (type 2 and 3) and others. As shown in Fig. 4-14*a*, a pulse-forming network is first needed which converts the incoming UP and DN pulses into a counting clock and a direction ($\overline{\text{UP}}$/DN) signal (as explained by the waveforms in Fig. 4-14*b*). A practical pulse-forming circuit of this kind is shown in Fig. 7-7.

On each UP pulse generated by the phase detector, the content $N$ of the UP/DOWN counter is incremented by 1. A DOWN pulse will decrement $N$ in the same manner. The content $N$ is given by the $n$-bit parallel output signal $u_f$ of the loop filter. Because the content $N$ is the weighted sum of the UP and DN pulses—the UP pulses having an assigned weight of $+1$, the DN pulses, $-1$—this filter can roughly be considered an *integrator* having the transfer function

$$H(s) = \frac{1}{sT_i}$$

where $T_i$ is the integrator time constant. This is, however, a very crude approximation, since the UP and DN pulses do not carry any information (in this application at least) about the actual size of the phase error; they only tell whether the phase of $u_1$ is leading or lagging $u_2$.

A similar type of all-digital loop filter, that is also compatible with digital PDs such as the XOR or the $JK$ flip-flop, is the $K$ counter[18] as shown in Fig. 4-15a. The $K$ counter is an ordinary reversible counter; its scaling factor $K$ can usually be set by control signals ($K$ modulus control). The clock input is normally a square wave having a frequency of $2^M f_0$, where $f_0$ is the frequency of the reference signal. (As will be shown in Sec. 4-2-5, this clock signal is mostly derived from the DCO.) The output of the PD, an XOR in this example, controls the direction of counting (UP or DN).

If the system operates without any phase error, $u_1$ and $u_2$ are out of phase by exactly 90° and the output of the XOR is a symmetrical square wave. Hence the counter will count upward and downward during identical time intervals; if $K$ is made large enough, the counter will neither overflow nor underflow, provided that counting upward starts at a content

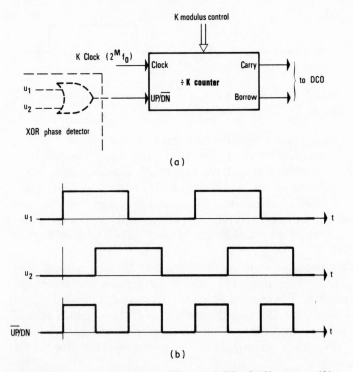

Fig. 4-15 *K*-counter loop filter. (*a*) Block diagram. (*b*) Corresponding waveforms.

of zero. (The condition of no overflow or underflow is met when $K$ is at least $2^M/4$.)[18]

If the signal $u_1$ starts leading the signal $u_2$, the XOR output signal becomes asymmetrical, and the counter will count upward for longer time intervals than it counts downward. As a consequence, the counter will overflow from time to time and generate a CARRY pulse. In the opposite case—$u_1$ lagging $u_2$—BORROW pulses will be produced by the counter. The CARRY and BORROW pulses can be used to control a DCO by adding or deleting pulses generated by a fixed-frequency oscillator on every CARRY and BORROW, respectively.[18] Alternatively, the CARRY and BORROW pulses could also be used to increment or decrement the scaling factor of another frequency divider whose clock input is also taken from a fixed-frequency oscillator.[19]

It is not easy to indicate a transfer function $H(s)$ for this type of loop filter, because its operation is nonlinear. At zero phase error, it does not produce any output signal at all. For a constant nonzero phase error it produces CARRY or BORROW pulses of *constant* frequency. If these pulses are used to add or delete pulses generated by a fixed-frequency oscillator, the $K$ counter appears as a constant-gain block rather than as a filter. If the CARRY and BORROW pulses are used to alter the scaling factor of a DOWN counter with fixed clock frequency, its operation can roughly be considered as an integrating action. Mathematically speaking, it is questionable whether this device can be considered as a filter at all.

Another device of this kind is the so-called $N$-before-$M$ filter (Fig. 4-16).[11] The performance of this filter is very nonlinear. In Fig. 4-16 it is suggested that the $N$-before-$M$ filter operates in conjunction with a phase detector generating UP and DN DOWN pulses, as was the case with the type 4 phase detector (Table 2-1). The $N$-before-$M$ filter uses two frequency counters scaling down the input signal by a factor $N$ and one counter scaling down by $M$, where $M > N$ always. The $\div M$ counter counts the incoming UP *and* DN pulses, as shown in Fig. 4-16. As also seen in the diagram, the upper $\div N$ counter will produce one CARRY output when it has received $N$ UP pulses. But it will generate this CARRY only when the $\div M$ counter does not receive $M$ pulses. Otherwise the $\div N$ counter would have been *reset*. We can say that the upper $\div N$ counter will produce a CARRY pulse whenever more than $N$ pulses of an ensemble of $M$ pulses have been UP pulses, or whenever a defined majority of incoming pulses are UP pulses. A similar statement can be made for the lower $\div N$ counter in Fig. 4-16, which will output BORROW pulses only when the majority of incoming pulses are DN pulses.

The outputs of the $N$-before-$M$ filter can be used in a similar way to control a DCO, as indicated for the $K$ counter.

We will now deal with digital loop filters compatible with an $N$-bit

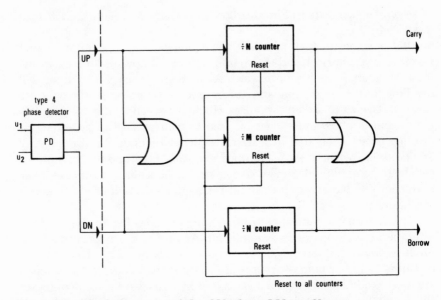

**Fig. 4-16   Block diagram of the *N*-before-*M* loop filter.**

parallel input signal. The obvious solution for this case is the *digital filter,* which operates by itself with *N*-bit input and *N*-bit output signals. With digital loop filters, any desired transfer function performed by an analog loop filter (and many additional ones) can be reproduced. Take, for example, the transfer function of the single-pole, low-pass type 1 filter (Table 2-2), which reads

$$F(s) = \frac{1}{1 + sT_1}$$

If such a filter has to be built as a digital filter, its impulse transfer function or *z* transform *F*(*z*) has to be found first, most likely from a *z*-transform table.[20,21] In the case of a first-order low-pass filter, the *z* transform reads

$$F(z) = \frac{T/T_1}{1 - z^{-1}e^{-T/T_1}} \tag{4-27}$$

where $z^{-1}$ is the delay operator corresponding to

$$z^{-1} = e^{-sT}$$

and *T* is the sampling period used for the digital filter.

Using Eq. (4-27), the $z$ transform $U_f(z)$ of the loop filter output signal $u_f(t)$ is given by

$$U_f(z) = U_d(z) \frac{T/T_1}{1 - z^{-1}e^{-T/T_1}} \tag{4-28}$$

$$= U_d(z) \frac{a}{1 - bz^{-1}}$$

where $U_d(z)$ is the $z$ transform of the phase-detector output signal $u_d(t)$. The filter coefficients have been expressed by $a$ and $b$ for simplicity, where $a = T/T_1$ and $b = \exp(-T/T_1)$.

Transforming Eq. (4-27) back into the time domain yields

$$u_f(nT) = au_d(nT) + bu_f[(n-1)T] \tag{4-29}$$

where $u_f(nT)$ signifies the sample of $u_f(t)$ taken at the discrete time $t = nT$.

Equation (4-29) is the equation of the recursive first-order, low-pass filter. It tells that the output signal of the loop filter at time $t = nT$ is calculated from the sample of the input signal $u_d(nT)$ taken at that time and from the output sample $u_f[(n-1)T]$ calculated in the previous sampling interval. The structure of the digital filter can now be drawn (Fig. 4-17). The operator $z^{-1}$ stands for a delay by one sampling interval $T$.

Digital filters can be implemented by hardware or by software. If implemented by hardware, the delay operation $z^{-1}$ is produced by a buffer register. If implemented by software, the delayed sample $u_f[(n-1)T]$ is stored either in an internal register (if any) of the central processing unit

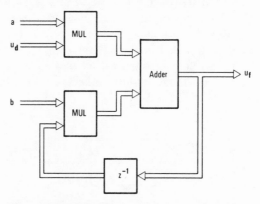

**Fig. 4-17  Block diagram of a loop filter fabricated from a recursive first-order digital filter.**

(CPU) or in a memory location. The final choice of a hardware or a software implementation depends primarily on the required speed of operation, as already pointed out by the introductory remarks in Sec. 4-2-1.

In any case, the sampling rate of the digital filter must be chosen appropriately. According to the Nyquist theorem, the sampling rate should be at least twice the bandwidth of the signal (at least twice the frequency $f_0$ of the signals $u_1$ and $u_2$ of the phase detector). To minimize aliasing effects,[17,21] the sampling rate is normally chosen to be at least 10 times the operating frequency $f_0$. Some applications of digital loop filters in ADPLLs will be given in Sec. 4-2-5.

### 4-2-4 Digital-Controlled Oscillators

A variety of DCOs can be designed; they can be implemented by hardware or by software. We will consider the most obvious solutions here.

Probably the simplest solution is the $\div N$ counter DCO (Fig. 4-18). A $\div N$ counter is used to scale down the signal generated by a high-frequency oscillator operating at a fixed frequency. The $N$-bit parallel output signal of a digital loop filter is used to control the scaling factor $N$ of the $\div N$ counter.

Another DCO type is the so-called *increment-decrement* (ID) counter shown in Fig. 4-19a.[18,22] This DCO is intended to operate in conjunction with those loop filters that generate CARRY and BORROW pulses, such as the $K$ counter or the $N$-before-$M$ filter discussed in Sec. 4-2-3. The operation of the ID counter follows from the waveforms shown in Fig. 4-19b. In the absence of CARRY and BORROW pulses, the ID counter divides the input frequency by a factor of 2. Whenever a CARRY pulse appears at the increment input, a clock cycle is added by the internal logic of the ID counter. A clock cycle is deleted on the other hand, whenever a BORROW pulse is applied to the decrement input. Thus the output frequency of the DCO can be controlled within a range given by the maximum frequency of the CARRY and BORROW pulses. A detailed analysis of this type of circuit is found in Ref. 18.

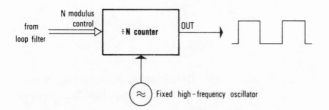

**Fig. 4-18** **Block diagram of a** $\div N$ **counter DCO.**

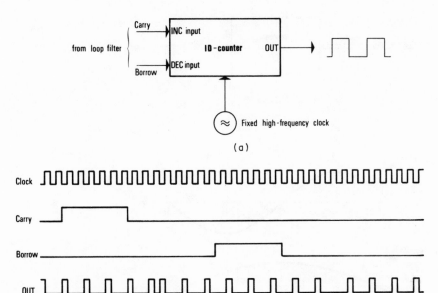

Fig. 4-19   Increment-decrement-counter DCO. (*a*) Block diagram. (*b*) Corresponding waveforms.

The two DCO circuits discussed earlier are better suited for hardware than for software implementations. The waveform synthesizer DCO—the third and last DCO to be considered here—lends itself almost ideally to implementation by software. This type of DCO generates sine and/or cosine waveforms by looking up tables stored in read only memory (ROM).[16] The block diagram of a waveform-synthesizer DCO is shown in Fig. 4-20a.

The waveforms in Fig. 4-20b demonstrate how the synthesizer generates sine waves of different frequencies (1 Hz and 0.5 Hz in this example). It operates at a fixed clock rate (i.e., it calculates a sample of the synthesized signal at the sampling instants $t = 0, T, 2T, \ldots, nT$, irrespective of the desired frequency). Lower-frequency signals are generated with higher resolution than higher-frequency signals.

In the example of Fig. 4-20b it has been arbitrarily assumed that the sampling period is 20 ms; most actual waveform synthesizers operate much faster, of course. It shows that a 1-Hz sine wave is generated with a resolution of 20 samples for a full period. In case of the 0.5-Hz sine wave, twice as many samples (40) are produced within a full period. When generating a 1-Hz sine wave, the synthesizer calculates the sine function for

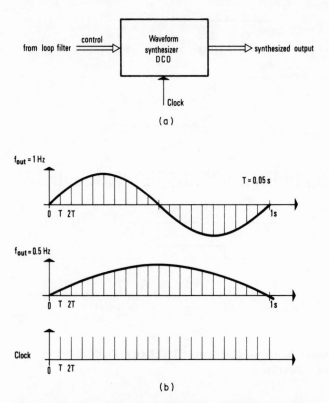

**Fig. 4-20  Waveform-synthesizer DCO. (*a*) Block diagram. (*b*) The waveforms show how sine waves with frequencies of 1 Hz and 0.5 Hz, respectively, are synthesized.**

the phase angles $\phi = 0$ (initial value), $2\pi/20$, $2(2\pi/20)$, $3(2\pi/20)$, . . . , and the phase angle $\phi$ is incremented by an amount $\Delta\phi = 2\pi/20$ at every clock period.

If an arbitrary frequency $f$ is to be produced, the increment $\Delta\phi$ is given by

$$\Delta\phi = 2\pi fT$$

where $T$ is the sampling period.

The synthesizer calculates $\Delta\phi$ from the $N$-bit parallel $f$-control signal delivered by the loop filter (shown in Fig. 4-20*a*). The waveform synthesizer is capable of generating the appropriate signal to the $f$-control input. It is no problem to generate "simultaneously" a sine and a cosine function, as required, when a Hilbert-transform phase detector is used with an ADPLL system (refer to Fig. 4-12*a*).

Digital waveform synthesizers are easily implemented using single-chip microcomputers.[16] The speed of trigonometric computations can be greatly enhanced by using table-lookup techniques rather than by calculating a sine function with a Taylor series or Chebysheff polynomials. An example of the application of a waveform-synthesizer DCO is presented in Sec. 4-2-5.

### 4-2-5 Examples of Implemented ADPLLs

Based on the numerous variants of all-digital PDs, loop filters, and DCOs, an almost unlimited number of ADPLLs can be built by combining compatible functional blocks. There is an extended literature on this subject, and a review of the most important systems is found in Ref. 11. A detailed discussion of every possible ADPLL system would go beyond the scope of this book; we will therefore consider only three typical ADPLL implementations.

The first two are hardware implementations. There is no reason why they could not be designed using software as well. The last example to be discussed is a typical software-based system, which encompasses a large variety of mathematical operations. A hardware implementation of this type of PLL is certainly not impossible, but the hardware would be very complex.

The first example of an ADPLL is depicted in Fig. 4-21$a$. This system is built from a flip-flop phase detector similar to type 3 (Table 2-1), an UP/DOWN-counter loop filter as previously shown in Fig. 4-14, and a $\div N$-counter DCO (already illustrated in Fig. 4-18). The pulse-forming network within the PD consists of the first $JK$ flip-flop ($JK$-FF$_1$) and an AND gate. The pulse-forming network generates the counting clock (CK) for the UP/DOWN counter of the loop filter. The second $JK$ flip-flop ($JK$-FF$_2$) is the intrinsic PD.

The waveforms in Fig. 4-21$b$ explain the operating principle of the PD. The $JK$-FF$_1$ scales down the input signal $u_1$ by a factor of 2; the scaled-down input signal is designated $u_1^*$. The reconstructed signal $u_2$ has the same frequency as $u_1^*$, not $u_1$. The waveforms in Fig. 4-21$b$ have been drawn for two cases:

1. $u_1^*$ leading $u_2$

2. $u_1^*$ lagging $u_2$

The $JK$-FF$_2$ is an edge-triggered flip-flop, as seen from the symbol in Fig. 4-21$a$. An edge-triggered $RS$ flip-flop could be used as well.

The positive transitions of $u_1^*$ set this flip-flop; the positive transitions

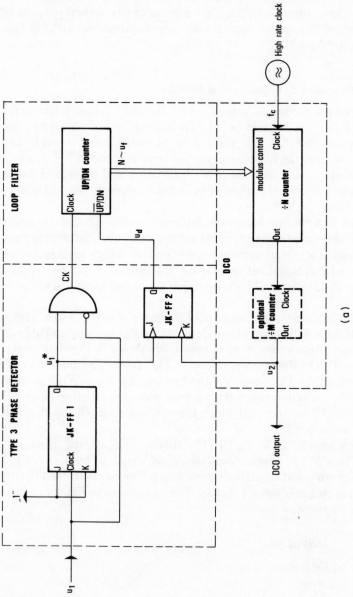

(a)

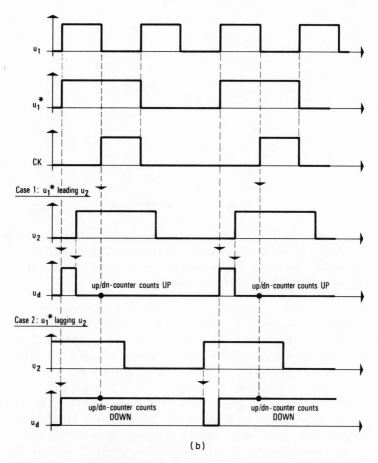

(b)

**Fig. 4-21   All-digital PLL system, example 1. (*a*) Left, block diagram. (*b*) Above, corresponding waveforms. Two cases are shown: (1) $u_1^*$ leading $u_2$; (2) $u_1^*$ lagging $u_2$.**

of $u_2$ reset it. As seen from the waveforms in the diagram, the counting clock CK for the UP/DOWN counter occurs at a time when the output signal $u_d$ or $JK$-FF$_2$ is stable, that is, high when $u_1^*$ lags $u_2$ or low when $u_1^*$ leads $u_2$. Consequently the phase-detector output signal $u_d$ is used as the direction input $\overline{\text{UP}}$/DOWN for the UP/DOWN counter. If the frequencies of $u_1^*$ and $u_2$ are not identical, the UP/DOWN counter will count upward or downward until $N$ has reached the value that causes the $\div N$ counter to generate the correct output frequency.

Because $N$ can only be varied in steps of 1, the frequency of the signal $u_2$ will normally be slightly too high or too low. This will force the contents of $N$ to jitter continuously around the values of $N$ and $N + 1$ if the ref-

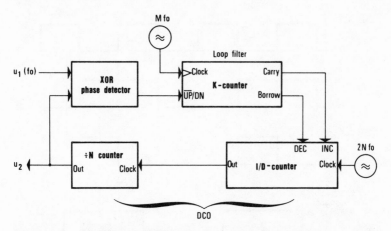

**Fig. 4-22 All-digital PLL system, example 2. This system is realized by the integrated circuit SN74LS297 manufactured by Texas Instruments. Note that many other configurations are possible when using this circuit.**

erence frequency $f_1$ is constant. At equilibrium $f_1$ will be equal to $f_c/(N \cdot M)$, where $M$ is the scaling factor of the optional $\div M$ counter.

The second ADPLL system is shown in Fig. 4-22. This system was integrated in LS-TTL technology (SN74LS297 manufactured by Texas Instruments) and can be configured in many variants.[18,22] Figure 4-22 shows one of many configurations. Furthermore, two different PDs are on the chip, an XOR (type 2) and a flip-flop PD (type 3). The loop filter used in this ADPLL is the previously discussed $K$-counter loop filter (see Fig. 4-15); the DCO, the ID counter DCO (see Fig. 4-19), has also been discussed previously.

The system of Fig. 4-22 is supposed to operate at a center frequency $f_0$ (referred to the input $u_1$). The $K$ counter and the ID counter are driven by clock signals having frequencies of $M$ times and $2N$ times the center frequency $f_0$, respectively. Normally both $M$ and $N$ are integral powers of 2, and are easily derived from the same oscillator.

If the input frequency is exactly $f_0$, the ID counter divides exactly by 2, and $u_1$ and $u_2$ have exactly the same frequency. The $K$ counter does not produce any CARRY or BORROW pulses, so the system is at equilibrium. If the input frequency is now increased, the $K$ counter starts generating CARRY pulses, which causes the ID counter to *add* pulses from time to time. The average number of pulses added every second is $N$ times the deviation $\Delta f$ of the input frequency from the center value $f_0$ (as is easily seen from Fig. 4-22). A detailed mathematical treatment of this ADPLL system is found in Ref. 22.

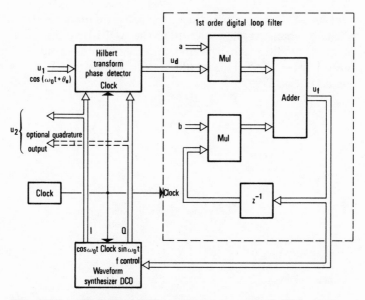

**Fig. 4-23   All-digital PLL system, example 3. This system is best implemented by software.**

The last example of an ADPLL is illustrated in Fig. 4-23. A similar system has been implemented by software on a TMS320 single-chip microcomputer (Texas Instruments).[16] The system of Fig. 4-23 is built from functional blocks introduced previously.

1. A Hilbert-transform phase detector (Fig. 4-12)

2. A first-order digital loop filter (Fig. 4-17)

3. A waveform-synthesizer DCO (Fig. 4-20)

As indicated in the block diagram, the arithmetic and logic operations within the functional blocks are performed under control of a clock. This means that all routines calculating the output variables of the blocks are executed periodically. The DCO generates the in-phase and quadrature signals $I$ and $Q$ required by the Hilbert-transform PD to calculate the phase error, $u_d \sim \theta_e$. The output signal $u_d$ is digitally filtered by the loop filter, which performs the operation

$$u_f(nT) = au_d(nT) + bu_f[(n-1)T] \tag{4-29}$$

where $a$ and $b$ are constants of the filter and have to be derived from the desired filter cutoff frequency and sampling rate as shown by Eq. (4-28).

One of the major benefits of the software implementation is the simplicity of changing the structure of the ADPLL system. With only minor program modifications the first-order loop filter could be turned into a second-order one.[16] This would yield a third-order PLL, as will be discussed in Sec. 7-9.

# 5 DESIGNING PLL CIRCUITS

The theories of linear and digital PLLs were discussed in Chaps. 3 and 4. It was shown that these two categories differ greatly from each other, and so the designer must first decide which type should be chosen as well as choose from a number of different types of PDs and loop filters. In this section, we will discuss the criteria leading to the designer's decisions. The results are summarized in a flow chart (Fig. 5-1).

## 5-1 SELECTING THE APPROPRIATE TYPE OF PLL

The designer of a PLL circuit will first have to decide whether a linear or a digital PLL should be built. This decision is influenced mainly by the type of reference signal and by the desired mode of operation of the PLL. If a PLL will be used as as a tracking filter, it should be capable of locking within a limited frequency band only; furthermore it should be able to filter out noise signals. Hence, the *linear* PLL is the only reasonable solution.

The situation is completely different, however, if the PLL will be used as a frequency synthesizer. Here the PLL should provide locking within a broad frequency range; hence, a *digital* PLL will be chosen. A DPLL will also generally be preferred if the reference signal is a square wave.

The final selection of a particular PLL IC or hybrid circuit will depend on other factors, such as the usable range of supply voltage(s), the question of unipolar (e.g., $+5$-V) or bipolar (e.g., $\pm 7.5$-V) power supplies, the compatibility of reference and output signals with logic families (TTL, CMOS, ECL), and, of course, the useful frequency range.

A number of PLL ICs, mainly built in $p$-channel or complementary metal-oxide semiconductor (PMOS or CMOS) technologies, can be used only at relatively low frequencies; TTL circuits, however, can operate at up to 25 MHz or more, and emitter-coupled logic (ECL) circuits cover the frequency range up to about 1 GHz.

## 5-2 SELECTION OF THE OPTIMUM PHASE DETECTOR

If the designer has decided to use a linear PLL, the only suitable PD is the type 1, as mentioned previously.

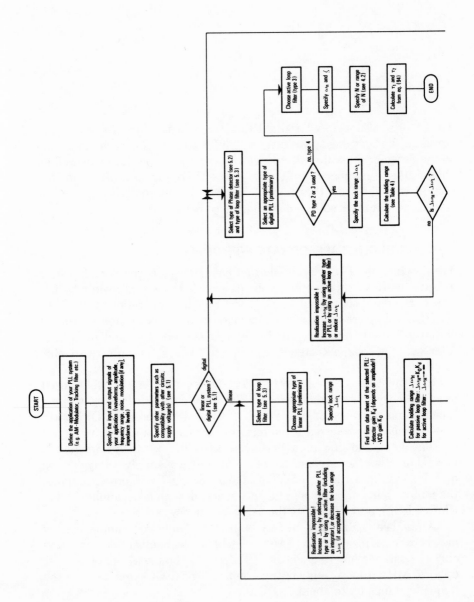

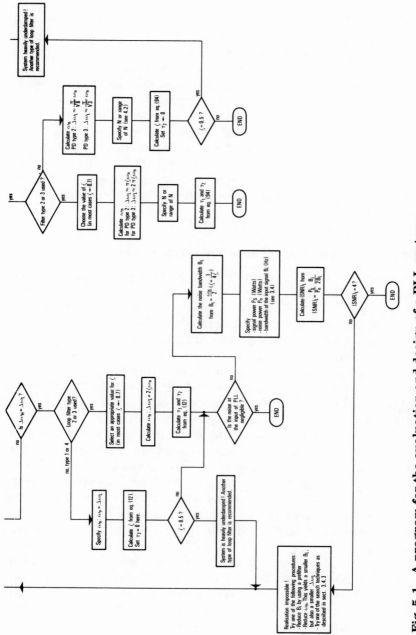

**Fig. 5-1  A program for the evaluation and design of a PLL system.**

113

If a DPLL has been chosen, a number of different PDs can be specified. The type 2 PD has properties comparable to those of the type 1 PD. Choosing this type will result in limited lock and hold ranges. When this type of PLL is operating at center frequency $\omega_0$, there will be a 90° phase offset between reference and output signals, as was the case with the linear PLL. We will have to remember that the performance of a type 2 PD depends on the duty-cycle ratios of the signals.

As already mentioned the performance of type 3 and type 4 PDs is independent of these duty-cycle ratios. It was shown in Sec. 4-1-1 that the type 4 PD outperforms the type 3 PD with regard to pull-in capabilities, and so the type 4 will be preferred in most digital PLL applications. If the type 4 PD is selected in combination with a charge pump (Table 2-1), the reference and output signals will be nearly in phase over the full operating range of frequencies.

### 5-3 SELECTION OF THE OPTIMUM TYPE LOOP FILTER

For a second-order PLL four different types of loop filters can be used, as shown by Table 2-2. Type 1 and type 4 loop filters have one single pole, while type 2 and type 3 filters are characterized by a pole and a zero. For type 1 and 4 one single time constant $\tau_1$ must be specified, but an additional time constant $\tau_2$ must be defined for types 2 and 3.

Remember that the tracking performance of a PLL is described by the parameters $\omega_n$ and $\zeta$. As shown by the substitutions in Eq. (3-6), $\omega_n$ and $\zeta$ are determined by $\tau_1$ and $\tau_2$. In most applications the designer wishes to specify $\omega_n$ and $\zeta$ independently from each other. Then $\omega_n$ could be given by a desired value of lock range or noise bandwidth, and $\zeta$ should probably take a suitable value of about 0.7 for optimum transient response. From Eq. (3-6) it is seen that $\omega_n$ and $\zeta$ can be chosen independently only if both time constants $\tau_1$ and $\tau_2$ are specified. If only one time constant, $\tau_1$, is defined, then it is possible to specify *either* $\omega_n$ *or* $\zeta$, the other parameter resulting as a consequence. In many cases where a relatively large time constant $\tau_1$ has been chosen, an unduly low damping factor $\zeta$ is obtained. For this reason a type 2 or type 3 loop filter will be preferred in most applications.

Finally, the designer will have to decide whether or not to use an active loop filter. If a digital PLL is built in combination with a type 4 PD, the active filter is preferred (as discussed in Chap. 4). For the remaining applications it is irrelevant whether an active or a passive loop filter is used. In most cases a passive filter is sufficient.

An active filter can present advantages in some special cases if a PLL is used to recover a signal that is not continuously available. An important example is the recovery of the color subcarrier in color television. There the reference signal is a burst having a low duty-cycle ratio. If an active loop filter is used, its output signal is "frozen" during the time when no reference signal exists.

Whenever an active filter is used for the loop filter, the bandwidth of the operational amplifier must be large enough to handle the high-frequency components of its input signal. In particular the type 3 loop filter is not only an integrator, but it also provides some "proportional gain" at higher frequencies.

## 5-4 A SIMPLE PROCEDURE FOR DESIGNING A PLL SYSTEM

Figure 5-1 is a flowchart for the practical design of a PLL. The procedure is written in the form of a computer program and consists of procedure blocks (squares) and decision blocks (diamonds). The circuit design begins at START and is completed when an END is reached. Such a program is of course not a fail-safe tool for designing PLLs of any complexity, but it does help the designer to ask some crucial questions.

# 6 STATE OF THE ART OF COMMERCIAL PLL INTEGRATED CIRCUITS (FALL 1983)

## 6-1 AVAILABLE PLL INTEGRATED CIRCUITS

A wide variety of ICs for PLL systems or parts thereof are available from semiconductor manufacturers. Table 6-1 lists the ICs on the market as of 1983 together with a short description of each chip. It indicates whether the circuit is a digital or linear PLL, the technology from which it is built (TTL, CMOS, and so on), and the other types of circuits with which it is compatible. The table also lists the special features of these ICs, such as additional operational amplifiers on the chip.

Table 6-1 shows that many chips contain complete PLL systems, but on others only the functional blocks of a PLL (such as a VCO or a PD) are implemented. At present fully integrated PLLs on a single chip operate at frequencies of up to 35 MHz. Higher-frequency ranges are easily obtained by combining various ICs containing functional blocks only.

There are two different types of fully integrated PLL systems on a single chip. In the first and larger group, all connections between the individual functional blocks are made internally on the chip. Thus the number of external components and connections is reduced, but the user has only limited freedom of individual design. If the output of the PD is directly connected to an input terminal of the VCO, the designer can use only a passive loop filter, because only one pin is left for hooking up an *RC* combination. There is a second group of fully integrated PLL ICs in which the individual functional blocks are uncommitted. This enables the designer to build individually tailored systems.

Table 6-1 also lists some specialized versions of PLLs. One of these special circuits is the so-called *Touch-Tone decoder*. A Touch-Tone decoder is merely a conventional PLL equipped with an additional in-lock detector (Fig. 6-1). Besides the usual PLL blocks, there are some additional circuits on the chip, such as a 90° phase shifter, an analog multiplier, an output filter, and a comparator (or Schmitt trigger).

The phase-shifted signal (often called the *quadrature* signal) can be taken directly from the VCO in most cases. As already described in Sec. 3-4-3, the output signal of the comparator is a logic 1 if the PLL is locked. To prevent false triggering, an output filter is used whose time constant can be made arbitrarily long by an external capacitor.

Text continues on page 144.

117

**Table 6-1.** *Commercially Available PLL ICs and Related Components, Fall 1983*

| Manufacturer | Part no. | Identification | Frequency range | Supply voltage(s) |
|---|---|---|---|---|
| Consumer Micro-circuits Ltd. | FX-003 | Multiple-tone decoder | 459 Hz–2.8 kHz | 4.5–6 V |
| | FX-101L | Single-tone switch | 1 Hz-20 kHz | 8–15 V |
| | FX-105 | Digital filter switch | 60 Hz–5 kHz | −10 V to −15 V |
| | FX-301L | High/low fre-quency switch | 1 Hz–20 kHz | 8–15 V |
| | FX-307 | 3-tone multicode detector | 100 Hz–7 kHz | 8–15 V |
| | FX-401 | Three-level fre-quency switch | 1 Hz–20 kHz | 8–15 V |
| | FX-501 | Bistable tone switch | 10 Hz–20 kHz | 8–15 V |
| | FX-601 | Monostable tone switch | 10 Hz–20 kHz | 8–15 V |

| Type of reference signal | Input/Output signal compatibility | Type of PD | Logic family | Comments |
|---|---|---|---|---|
| 35-mV sensitivity | Output signal = 4-bit word | Linear | CMOS | Detects 1 of 15 tones. Output is 4-bit digital word. |
| Sine or square wave | Hi/lo | Zero-crossing detector | PMOS | Zero-crossing detection technique enables detection of signals whose frequency is either within a band $f_1$ to $f_2$ (band mode) or is higher than a preset value $f_1$ (datum mode). For low-noise applications only. |
| Asymmetrical | Output NPN, open collector | Special | PMOS | Special PLL circuit with integrated Touch-Tone decoder. Only tone-decoder output available. |
| 35-mV sensitivity | Hi/lo | Zero-crossing detector | PMOS | Similar to FX-101L, but also enables datum switching with adjustable hysteresis. |
| 35-mV sensitivity | 3-bit word | Zero-crossing detector | PMOS | Three output signals indicating which tone of 3 programmable tones has been detected. |
| 35-mV sensitivity | Inband, hi, lo | Zero-crossing detector | PMOS | Three binary outputs indicating whether the received frequency is below, within, or above a selected frequency band. |
| 18-mV sensitivity | See comment | Zero-crossing detector | PMOS | Output is bistable (toggle-FF). First tone-burst detector within selected frequency band sets the FF, next tone within band resets it, etc. |
| 18-mV sensitivity | See comment | Zero-crossing detector | PMOS | Similar to FX-501, but output is monostable. Each tone burst detected within selected frequency band triggers a one-shot oscillator. |

| Manufacturer | Part no. | Identification | Frequency range | Supply voltage(s) |
|---|---|---|---|---|
| Consumer Microcircuits Ltd. | FX-701P | Trilevel frequency switch | 1 Hz–10 kHz | 8–15 V |
| | FX-205 | Tone encoder | 0–50 kHz | 10–15 V |
| | FX-207 | Three-tone multicode encoder | 100 Hz–7 kHz | 8–15 V |
| | FX-503 | Multitone encoder | 459 Hz–2.8 kHz | 4.5–6 V |
| | FX-107 | Three-tone encoder/decoder | 100 Hz–7 kHz | 8–15 V |
| | FX-407 FX-507 | Five-tone sequential code transceiver | 1–3 kHz | −10 V to −15 V |
| | FX-607 | Five-tone encoder/decoder | 631 Hz–1.8 kHz | 10–15 V |
| | FX-4070/4071 | Hybrid 5-tone encoder/decoder | | |
| | FX5070/5071 | Hybrid 5-tone encoder/decoder | | |
| Exar | XR-210 | FSK modulator/ demodulator | 0–20 MHz | 5–26 V |

| Type of reference signal | Input/Output signal compatibility | Type of PD | Logic family | Comments |
|---|---|---|---|---|
| 35-mV sensitivity | Inband, hi, lo | Zero-crossing detector | PMOS | Detects whether a received signal's frequency is below, within, or above a selected frequency band. |
| — | — | — | PMOS | Generates staircase approximated sine wave (8 levels). |
| — | Square wave | — | PMOS | Generates a sequence of three tones, 8 permutations available on control of three signals *RA*, *RB*, *RC*. |
| — | — | — | CMOS | Output is staircase approximated sine wave (16 levels). Refer to data sheet for further information |
| 35-mV sensitivity | Output = square wave | — | PMOS | Three-tone sequential encoder and decoder with automatic transponding capability. |
| Asymmetrical | Output PMOS | Special | PMOS | Special PLL circuit, 5-tone decoder. Output signal of VCO is a staircase approximation of a sine wave. |
| 50-mV sensitivity | See comment | — | PMOS | Similar to FX-407, FX-507. |
| | | | Thick-film hybrid | Hybrid versions of FX-407 including most of the external components on chip. |
| | | | Thick-film hybrid | Hybrid versions of FX-507 including most of the external components on chip. |
| Asymmetrical | Output NPN, open collector | Linear | Bipolar | Special PLL circuit with Touch-Tone decoder output. Application: FSK transmitter/receiver. Additional control inputs: VCO sweep input, VCO gain control, VCO keying input, VCO fine tuning. |

**Table 6-1. Commercially Available PLL ICs and Related Components, Fall 1983 (Continued)**

| Manufacturer | Part no. | Identification | Frequency range | Supply voltage(s) |
|---|---|---|---|---|
| Exar | XR-215 | PLL | 0–35 MHz | 5–26 V |
| | XR-567 | Tone decoder | 0–500 kHz | +5 V |
| | XR-S200 | Multifunction IC | 0–30 MHz | ±10 V (±3 to ±30 V) |
| | XR-2206C | Monolithic function generator | 0–1 MHz | 10–26 V |
| | XR-2207 | VCO | 0–1 MHz | 8–26 V |
| | XR-2208 | Operational multiplier | 0–100 MHz (transconductance bandwidth) | ± 4.5 V to ± 16 V |
| | XR-2213 | Single-tone decoder | 0.01 Hz–300 kHz | |
| Harris | HA-2800/2805 | PLL | 0–25 MHz | + 5 V and − 15 V |
| | HA-2820/2825 | PLL | 0–3 MHz | ±6 V |
| Intersil | ICM 7240 | Programmable counter/timer | 0–13 MHz (15 V) | 2–16 V |
| | ICM7250 | Programmable counter/timer | 0–13 MHz (15 V) | 2–16 V |
| | ICM 7260 | Programmable counter/timer | 0–13 MHz (15 V) | 2–16 V |

| Type of reference signal | Input/Output signal compatibility | Type of PD | Logic family | Comments |
|---|---|---|---|---|
| Asymmetrical | DTL, TTL, ECL | Linear | Bipolar | Integrated op amp (output buffer), VCO sweep input, VCO gain control, range select pin. |
| Asymmetrical | TTL logic output 100 mA/20V | Linear | Bipolar | Special circuit (Touch-Tone decoder), receiver only. |
| Asymmetrical | — | Linear | Bipolar | Multipurpose circuit, contains multiplier, op amp, VCO, no internal connections. |
| — | TTL | — | Bipolar | VCO with square-wave and triangular signal output. FSK input selects 1 of 2 timing resistors. |
| — | TTL (keying inputs) | — | Bipolar | Two binary keying inputs select 1 or 2 of 4 timing resistors. |
| Asymmetrical | — | — | Bipolar | Four-quadrant multiplier + op amp + buffer amplifier (unity gain) |
| | | | | Chip includes preamp, VCO, PD, quadrature detector, voltage comparator. |
| Differential | TTL/ECL | Linear | Bipolar | Output signal TTL or ECL compatible, selected by "level control" input. 2 independent VCO outputs. On-chip lock-range limiter. |
| Differential | TTL | Linear | Bipolar | |
| — | CMOS/TTL | — | CMOS | Scaling factor (when used as a frequency divider) 1–255. |
| — | CMOS/TTL | — | CMOS | Scaling factor (when used as a frequency divider) 1–99. |
| — | CMOS/TTL | — | CMOS | Scaling factor (when used as a frequency divider) 1–59. |

| Manufacturer | Part no. | Identification | Frequency range | Supply voltage(s) |
|---|---|---|---|---|
| Intersil | 8038 | Precision waveform generator/VCO | 0–1 MHz | 5–28 V |
| Motorola | MC 4324/4024 | Dual-VCO | 0–25 MHz | +5 V |
| | MC 1648 | VCO | 0–225 MHz | +5.0 V or −5.2 V |
| | MC 4344/4044 | PD | 0–25 MHz | +5 V |
| | MC 12000 | Digital mixer/translator | 0–250 MHz | +5.0 V or −5.2 V |
| | MC 12009 | Dual-modulus prescaler | 600 MHz | +5 V or −5.2 V |
| | MC 12011 | Dual-modulus prescaler | 600 MHz | +5 V or −5.2 V |
| | MC 12012 | 2-modulus prescaler | 0–200 MHz | +5.0 V or −5.2 V |
| | MC 12013 | Dual-modulus prescaler | 600 MHz | +5 V or −5.2 V |
| | MC 12014 | Counter control logic | 0–25 MHz | 5 V |
| | MC 12040 | PD | 0–80 MHz | +5.0 V or −5.2 V |
| | MC 74416 MC 74417 MC 74418 MC 74419 | Programmable modulo-$N$ counters | 0–8 MHz min. | +5 V |
| | MC 14568 | Phase comparator and programmable counters | Up to 10 MHz | 3–18 V |

| Type of reference signal | Input/Output signal compatibility | Type of PD | Logic family | Comments |
|---|---|---|---|---|
| — | TTL | — | Bipolar | Generates simultaneously square, triangular, and sine-wave signals. |
| — | TTL | — | TTL | |
| — | ECL | — | ECL | |
| Asymmetrical | TTL | Digital types 2 and 4 | | Two different phase detectors on chip. |
| — | TTL/ECL | — | ECL | Refer to Sec 7-7, TTL/ECL and ECL/TTL translator on chip. |
| — | TTL/ECL | — | | Original divide values $\div 5/6$ expandable by external control signals to $\div 80/81$ and $\div 64/65$. |
| — | TTL/ECL | — | | Original divide values $\div 8/9$ expandable by external control signals to $\div 128/129$ and $\div 64/65$. |
| — | ECL/TTL | — | ECL | Refer to Sec. 7-7, TTL/ECL and ECL/TTL translator on chip. |
| — | TTL/ELC | — | | Original divide values $\div 10/11$ expandable by external control signals to $\div 160/161$ and $\div 80/81$. |
| — | TTL | — | TTL | Refer to Ref. 30. |
| Asymmetrical | ECL | Digital type 4 | ECL | |
| | TTL | — | TTL | Refer to Sec. 7-7. |
| Asymmetrical | CMOS/TTL | Type 4 | CMOS | Chip contains PD, a divide-by-4, -16, -64, or -100 counter, and a $\div N$ 4-bit programmable counter. |

| Manufacturer | Part no. | Identification | Frequency range | Supply voltage(s) |
|---|---|---|---|---|
| Motorola | MC 145143 | PLL frequency synthesizer | Up to 25 MHz | 4.5–12 V |
| | MC 145146 | PLL frequency synthesizer | 15 MHz max (500 mV$_{pp}$ input) | 3–9 V |
| | MC 145151 | Parallel input PLL frequency synthesizer | >30 MHz (5 V) | 3–9 V |
| | MC 145152 | PLL frequency synthesizer | 30 MHz | 3–9 V |

| Type of reference signal | Input/Output signal compatibility | Type of PD | Logic family | Comments |
|---|---|---|---|---|
| — | CMOS/ LSTTL | ? | CMOS | Chip contains PD and programmable frequency divider. Scaling factor is 32 to 8176 in steps of 16. |
| CMOS or low-level | CMOS | Type 4 | CMOS | Chip contains programmable reference (scaling factor 3–4096), oscillator, phase/frequency detector, in-lock detector, 10-bit programmable $\div N$ counter, programmable $\div A$ counter, control signal for (external) dual-modulus prescaler. Programming is performed over 4-bit data bus. |
| — | CMOS/ LSTTL | Type 4 | CMOS | Chip contains PD and programmable reference divider (Scaling factors: 8, 128, 256, 512, 1024, 2048, 2410, or 8196) and $\div N$ divider with scaling factor $3-16{,}383$ in steps of 1. Scaling factor $N$ is digitally controlled by parallel input port (14 bits wide). |
| CMOS or low-level | CMOS | Type 4 | CMOS | Chip contains programmable reference divider, 1 of 8 scaling factors (8, 64, 128, 256, 512, 1024, 1160, 2048) is selected by 3 divide select pins ($RA_0$, $RA_1$, $RA_2$), oscillator, phase/frequency detector, in-lock detector, 10-bit programmable $\div N$ counter, 6-bit programmable $\div A$ counter, control signal for (external) 2-modulus prescaler, fully-parallel programming. |

*Table 6-1. Commercially Available PLL ICs and Related Components, Fall 1983 (Continued)*

| Manufacturer | Part no. | Identification | Frequency range | Supply voltage(s) |
|---|---|---|---|---|
| Motorola | MC 145155 | Serial-input PLL frequency synthesizer | >30 MHz (5 V) | 3–9 V |
| | MC 145156 | Serial-input PLL frequency synthesizer | >30 MHz (5 V) | 3–9 V |
| | MC 145158 | PLL frequency synthesizer | | 3–9 V |
| | MC 145159 | PLL frequency synthesizer | | 3–9 V |

| Type of reference signal | Input/Output signal compatibility | Type of PD | Logic family | Comments |
| --- | --- | --- | --- | --- |
| — | CMOS/ LSTTL | Type 4 | CMOS | Chip contains PD and programmable reference divider (scaling factors: 16, 512, 1024, 2048, 3668, 4096, 6144, or 8192) and $\div N$ divider with scaling factor of 3–16383 in steps of 1. Scaling factor $N$ is digitally controlled by serial input signal (14 bits). 14-bit data latch integrated on chip. |
| — | CMOS/ LSTTL | Type 4 | CMOS | Chip contains PD and programmable reference divider (scaling factors: 8, 64, 128, 256, 640, 1000, 1024, or 2048) and programmable $\div N$ divider ($N = 3$–1023 in steps of 1) and $\div A$ divider ($A = 0$–127). Digital programming of scaling factors $A$ and $N$ by serial input signal $(10 + 7$ bits). An additional "modulus control" signal is used to control a 2-modulus prescaler (refer to Sec. 7-7). |
| | CMOS | Type 4 | CMOS | Chip contains programmable reference divider (scaling factor 3–16384), oscillator, phase/frequency detector, in-lock detector, 10-bit programmable $\div N$ counter, programmable $\div A$ counter, control signal for (external) 2-modulus prescaler. Programming by serial bit stream, data held in shift register/latch. |
| | CMOS | Type 4 | CMOS | Same specifications as MC 145158, but phase detector has differing output signal specs. Refer to data sheet. |

**Table 6-1.** *Commercially Available PLL ICs and Related Components, Fall 1983 (Continued)*

| Manufacturer | Part no. | Identification | Frequency range | Supply voltage(s) |
|---|---|---|---|---|
| Motorola | MC 146805T2 | Eight-bit micro-computer unit with PLL logic | 16 MHz max ($f_{IN}$-input) | 5 V |
| National Semiconductors, Ltd. | LM 565 | PLL | 0–1 MHz | $\pm 5$ to $\pm 12$ V |
| | LM 566 | VCO | 0–1 MHz | 10–24 V |
| | LM 1310 | PLL stereo demodulator | (38 kHz) | 10–18 V |
| | LM 1800 | PLL stereo demodulator | (38 kHz) | 12 V, 18 V max |
| | DS 8906 | AM/FM DPLL synthesizer | AM: 0.4–8 MHz FM: 60–120 MHz | 5 V |
| | MM 55104 | PLL frequency synthesizer | (max 3 MHz at $f_{IN}$-input) | 4.5–5.5 V |
| | MM 55114 | PLL frequency synthesizer | (max. 3 MHz at $f_{IN}$-input) | 7–10 V |

| Type of reference signal | Input/Output signal compatibility | Type of PD | Logic family | Comments |
|---|---|---|---|---|
| TTL/CMOS | TTL/CMOS | Type 4 | CMOS | Frequency synthesizer system built from a single-chip microcomputer. Chip includes 2508 bytes of user ROM, 64 bytes of RAM, oscillator, PD, 10-stage task-programmable reference divider, 14-bit binary variable divider etc. Instructions set similar to 6800 $\mu$P. |
| Differential | TTL | Linear | Bipolar | |
| — | — | — | Bipolar | Simultaneous square- and triangular-wave signals. |
| Asymmetrical | — | ? | Bipolar | Special circuit (stereo decoder) |
| Asymmetrical | — | ? | Bipolar | Special circuit (stereo decoder) |
| — | CMOS/TTL | Type 4? | ECL/I$^2$L | Chip contains oscillator (4-MHz), PD, charge pump, 120-MHz, 2-modulus prescaler, 20-bit shift register/latch for serial data input. |
| — | CMOS | Type 4? | CMOS | CB frequency synthesizer IC. Chip contains oscillator (normally 5.12 or 10.24 MHz), reference divider (selectable $\div 2^{10}$ or $\div 2^{11}$), PD, programmable frequency divider $\div N$ ($N = 1 - 225$) and in-lock detector. |
| — | CMOS | Type 4? | CMOS | Like MM 55104, with increased supply-voltage range. |

| Manufacturer | Part no. | Identification | Frequency range | Supply voltage(s) |
|---|---|---|---|---|
| National Semiconductors, Ltd. | MM 55106 | PLL frequency synthesizer | (max 3 MHz at $f_{IN}$-input) | 4.5–5.5 V |
| | MM 55116 | PLL frequency synthesizer | (max 3 MHz at $f_{IN}$-input) | 7–10 V |
| | MM 55107 | PLL frequency synthesizer | (max 3 MHz at $f_{IN}$-input) | 4.5–5.5 V |
| | MM 55108 | PLL frequency synthesizer with receive/transmit mode | (max 5 MHz at $f_{IN}$-input) (7.5-V supply) | 4.5–10 V |

| Type of reference signal | Input/Output signal compatibility | Type of PD | Logic family | Comments |
|---|---|---|---|---|
| — | CMOS | Type 4? | CMOS | CB frequency synthesizer chip. Contains oscillator (nominally 10.24-MHz), fixed reference prescaler $\div 2$, buffered oscillator output 5.12 MHz, programmable reference divider ($\div 2^9$ or $\div 2^{10}$), PD, programmable frequency divider $\div N$ ($N = 1$–$511$) and in-lock detector. |
| — | CMOS | Type 4? | CMOS | Like MM 55106, with increased supply-voltage range. |
| — | CMOS | Type 4? | CMOS | CB frequency synthesizer chip. Contains oscillator (nominal frequency 10.24-MHz) buffered 5.12-MHz output, selectable reference frequency divider ($\div 2^{10}$ or $\div 2^{11}$), PD, programmable frequency divider $\div N$ ($N = 1$–$255$) and in-lock detector. |
| — | CMOS | Type 4? | CMOS | CB frequency synthesizer chip. Contains oscillator (nominal frequency 10.24 MHz), fixed reference frequency divider $\div 2^{11}$, programmable frequency divider $\div N$ ($N = 1$–$511$). Output frequencies have a channel spacing of 5 kHz. In transmit mode, $N$ is incremented by 91, which corresponds to the IF frequency of 455 kHz. |

| Manufacturer | Part no. | Identification | Frequency range | Supply voltage(s) |
|---|---|---|---|---|
| National Semicondductors, Ltd. | MM 55110 | PLL frequency synthesizer with receive/transmit mode | (max 5 MHz at $f_{IN}$-input) (7.5-V supply) | 4.5–10 V |
| | MM 55109 | PLL frequency synthesizer and channel programmer | (max 3 MHz at $f_{IN}$-input) | 8–12 V |
| | MM 55111 | PLL frequency synthesizer and channel programmer | (max 3 MHz at $f_{IN}$-input) | 8–12 V |
| | MM 55117/ 18/19/20 | AM/FM/CB/SW PLL frequency synthesizer | (max 6 MHz at $f_{IN}$-input) | 6.5–10 V |

| Type of reference signal | Input/Output signal compatibility | Type of PD | Logic family | Comments |
|---|---|---|---|---|
| — | CMOS | Type 4? | CMOS | Like MM 55108, but the reference frequency divider has a selectable scaling factor of $2^{10}$ or $2^{11}$ and the scaling factor of the other frequency divider is programmable in the range $N = 1$–$1023$. |
| — | CMOS | Type 4? | CMOS | CB frequency synthesizer chip. Contains oscillator (nominal frequency 10.24 MHz), fixed reference divider $\div 2^{11}$ (channel spacing is 5 kHz), channel programmer for 40 channels with automatic presetting of the programmable frequency divider. Integrated channel number counter and drivers for 2-digit LED display. |
| — | CMOS | Type 4? | CMOS | Like MM 55109, but driver controls a fluorescent display. |
| — | CMOS | Type 4? | CMOS | 4 synthesizer ICs for CB or other transceivers MM 55117: 1 kHz/AM; 1 kHz/FM MM 55118: 5 kHz/CB; 5 kHz/SW MM 55119: 10 kHz/AM; 8 kHz/FM MM 55120: 5 kHz/AM; 4 kHz/FM |

| Manufacturer | Part no. | Identification | Frequency range | Supply voltage(s) |
|---|---|---|---|---|
| National Semiconductors, Ltd. | MM 55121 | Serial-data PLL frequency synthesizer | 0.1–4.5 MHz (reference input) | 6.5–10 V |
| | MM 55122 | Serial-data PLL frequency synthesizer | (max 4 MHz at $f_{IN}$-input) | 7–10 V |
| | MM 55123 | Serial-data PLL frequency synthesizer | — | — |
| | MM 55124 | PLL frequency synthesizer | (>3 MHz at $f_{IN}$-input) | 4.5–10 V |
| | MM 55126 | PLL frequency synthesizer | (>3 MHz at $f_{IN}$-input) | 4.5–10 V |

| Type of reference signal | Input/Output signal compatibility | Type of PD | Logic family | Comments |
|---|---|---|---|---|
| CMOS or low-level (min 0.8 $V_{pp}$) capacitively coupled | | Type 4 | CMOS | Chip contains oscillator (nominal frequency 10.24 MHz), fixed $\div 2^{11}$ reference divider, a 13-bit programmable frequency divider, a 15-bit serial-in shift register storing the scaling factor (13-bit) $+$ 2 control bits, a phase/fequency detector, buffered outputs 10.24 MHz, 320 kHz, 300 Hz, 60 Hz, channel spacing is 1 kHz. |
| — | CMOS | Type 4? | CMOS | CB frequency synthesizer chip, contains oscillator (nominal frequency 10.24 MHz), reference frequency divider $\div 2^{10}$, phase detector, programmable frequency divider $\div N$ ($N = 1-511$) and 3 4-bit D/A-converters for controlling squelch, AVC (automatic volume control), volume etc. Serial data input for chip programming. |
| | | | CMOS | Same specifications as MM 55121. |
| — | CMOS | Type 4? | CMOS | CB frequency synthesizer chip. Contains oscillator (nominal frequency 10.24 MHz), reference divider ($\div 2^{10}$ or $\div 2^{11}$), PD, programmable frequency divider $\div N$ ($N = 1-255$) and in-lock detector. |
| — | CMOS | Type 4? | CMOS | Similar to MM 55124, but has additional buffered 5.12-MHz output. $N$ is programmable in the range $N = 1-511$. |

*Table 6-1. Commercially Available PLL ICs and Related Components, Fall 1983 (Continued)*

| Manufacturer | Part no. | Identification | Frequency range | Supply voltage(s) |
|---|---|---|---|---|
| National Semiconductors, Ltd. | DS 8906 | AM/FM DPLL synthesizer | Up to 120 MHz | 5 V |
| | DS 8908 | AM/FM DPLL synthesizer | Up to 120 MHz | |
| Plessey | SL 650 | PLL | 0–500 kHz | $\pm 6$ V<br>$\pm 7.5$ V max |
| | SL 651 | | | |
| | SP 8792 | Programmable divider | min 200 MHz | 6.8–13.5 V or 5.2 V |
| | SP 8793 | Programmable divider | min 200 MHz | 6.8–13.5 V or 5.2 V |
| | NJ 8811 | Control circuit for frequency synthesis | max 2.5 MHz (counter input) max 10 MHz (oscillator input) | 5 V nom |
| | NJ 8812 | Control circuit for frequency synthesis | max 5 MHz (counter input) max 10 MHz (oscillator input) | 5 V |

| Type of reference signal | Input/Output signal compatibility | Type of PD | Logic family | Comments |
|---|---|---|---|---|
| Low-level (min 20 mV rms) | CMOS | Type 4 | (Mixed) | Circuit for AM/FM radios. Contains oscillator, PD, charge pump, 120-MGz ECL/$I^2L$ dual-modulus prescaler, 20-bit shift register/latch for serial data entry. |
| | | Type 4 | (Mixed) | Circuit for AM/FM radios. Contains oscillator, PD charge pump, operational amplifier, 120-MH$_Z$ ECL/$I^2L$ dual-modulus prescaler, 19-bit shift register/latch for serial data entry. |
| Differential | — | Linear | Bipolar | Additional op amp included, VCO gain control, 2 binary keying inputs for selecting 1 or 2 timing resistors. |
| | | | | Similar to SL 650, but without additional op amp. |
| Capacitive coupling | TTL/CMOS | — | ? | 2-modulus prescaler, scaling factors 80/81. |
| Capacitive coupling | TTL/CMOS | — | ? | 2-modulus prescaler, scaling factors 40/41. |
| — | NMOS | Type 4? | NMOS | Chip contains reference frequency divider with 16 selectable scaling factors, 2 programmable 4-bit counters for controlling 2 4-modulus prescalers (these are not included on chip), a programmable 8-bit divider and a PD. |
| — | | Type 4? | NMOS | Chip contains a reference frequency divider having 16 selectable scaling factors, one 6-bit programmable divider for controlling a 2-modulus prescaler (not integrated on chip), a programmable 8-bit divider and a PD. |

| Manufacturer | Part no. | Identification | Frequency range | Supply voltage(s) |
|---|---|---|---|---|
| Plessey | SP 8901 | 4-modulus divider (prescaler) | min 1 GHz | 5 V nom |
| | SP 8906 | 4-modulus divider (prescaler) | min 500 MHz | 5 V nom |
| RCA | CD 4046 | PLL | 0–1.2 MHz | 5–15 V |
| Signetics/Philips | LN 123 (HEF 4750) | Frequency synthesizer | 15 MHz (10 V) | 10 V nom |
| | LN 124 (HEF 4751) | Universal divider | >9 MHz (10 V) | 10 V nom |
| | NE 560 | PLL | 0–30 MHz | 16–26 V |
| | NE 561 | PLL | 0–30 MHz | 16–26 V |
| | Ne 562 | PLL | 0–30 MHz | 16–26 V |
| | NE 564 | PLL | 50 MHz | 5 V |
| | NE 565 SE 565 | PLL | 0–500 kHz | ±5 to ±12 V |

140

| Type of reference signal | Input/Output signal compatibility | Type of PD | Logic family | Comments |
|---|---|---|---|---|
| — | TTL/MOS | — | ? | Two binary control signals select a scaling factor of 478, 480, 510, or 512. |
| — | TTL/MOS | — | ? | Two binary control signals select a scaling factor of 239, 240, 255, or 256. |
| CMOS | | Digital types 2 and 4 | CMOS | Two different PDs source follower for $u_f$ signal included. Asymmetrical lock range can be chosen. |
| — | CMOS | Two PDs, special types | CMOS | Chip includes oscillator, reference divider with selectable scaling factor of 1, 2, 10, or 100, programmable frequency divider $\div N$ ($N = 1$–1023), 2PDs, modulator for FM or PM. |
| — | CMOS | — | CMOS | Refer to data sheet HEF 4751 |
| Differential | — | Linear | Bipolar | Tracking range control pin enables controlling the lock range $\Delta\omega_L$. VCO fine-tuning input, de-emphasis pin for FM detector, DC offset control pin for FM output. |
| Differential | — | Linear | Bipolar | Similar to NE 560, contains additional multiplier for use as a quadrature detector (AM demodulator). |
| Differential | — | Linear | Bipolar | Similar to NE 560, with additional buffer amplifier for the VCO output. |
| Asymmetrical | TTL | Type 1 | ? | Post-detection processor and limiting circuit at the input of the PD integrated on chip. |
| Differential | DTL/TTL | Linear | Bipolar | |

| Manufacturer | Part no. | Identification | Frequency range | Supply voltage(s) |
|---|---|---|---|---|
| Signetics/Philips | NE 566<br>SE 566 | VCO (function generator) | 0–500 kHz | 10–24 V |
| | NE 567<br>SE 567 | Tone decoder | 0–500 kHz | 5 V, 10 V max |
| | TDA 1005 | PLL stereo decoder | (38 kHz) | 8–16.5 V |
| SSS | SCL 4046 | PLL | 0–5 MHz | |
| | SCL 4446B | PLL | Improved version of SCL 4044 | |
| Texas Instruments | SN74LS124 | Dual VCO | 0–35 MHz guar.<br>50 MHz typ. | 5 V |
| | SN74S124 | Dual VCO | 0–60 MHz guar.<br>85 MHz typ. | 5 V |
| | SN74S297N | Digital PLL | 0–50 MHz | 5 V |
| | SN74LS324 | VCO | 0–30 MHz | 5 V |
| | SN47LS325<br>SN74LS326<br>SN74LS327 | Dual VCO | 0–30 MHz | 5 V |
| | SN74LS624<br>to SN4LS627 | Improved versions to types LS324 to LS327 | | |
| | SN74LS628 | Similar to type LS324 | | |
| | SN74LS629 | Similar to type LS124 | | |
| | SN76115N | PLL stereo decoder | (38 kHz) | 8–16 V |

| Type of reference signal | Input/Output signal compatibility | Type of PD | Logic family | Comments |
|---|---|---|---|---|
| — | — | — | Bipolar | Square- and triangular-wave signals simultaneously available. |
| Asymmetrical | TTL (100 mA, 15 V) | Linear | Bipolar | Special circuit: Touch-Tone decoder. |
| Asymmetrical | — | ? | Bipolar | Stereo decoder circuit. |
| Asymmetrical | CMOS | Digital types 2 and 4 | CMOS | |
| — | — | — | CMOS | |
| — | TTL | — | LS-TTL | A PLL can be realized in combination with other ICs. |
| — | TTL | — | S-TTL | Refer to data sheet. |
| Asymmetrical | TTL | Types 2 and 3 | LS-TTL | ADPLL; refer also to Sec. 4-2-5. |
| — | LSTTL | — | TTL | |
| — | LSTTL | — | TTL | |
| | | | TTL | |
| — | — | — | TTL | |
| — | — | — | TTL | |
| — | — | — | ? | Pin compatible with MC 1310. |

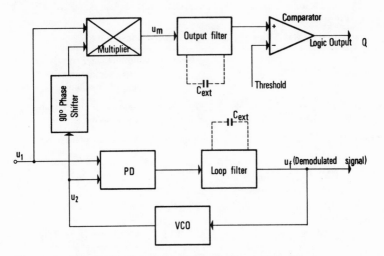

**Fig. 6-1   Block diagram of a Touch-Tone decoder.**

In many Touch-Tone decoder applications only the output of the in-lock detector is used, as in very simple telemetry systems where a binary 1 is characterized by the presence of a carrier and a 0 is given by no signal at all. In some specialized Touch-Tone decoder ICs, only the output signal of the in-lock detector is provided, while the $u_f$ signal (the output signal of the loop filter) is not available.

A successor of the Touch-Tone decoder is the *multitone decoder*. A multitone decoder is a special PLL system that outputs a logic 1 once a sequence of tones having different frequencies has been detected. In a particular IC five sequential tones are used. Each tone can have one of ten given frequencies. If repetitions of the same frequency are allowed (in some systems this is prohibited), $10^5$ different combinations are possible. Multitone decoders are very often employed in mobile radio station networks where one particular station has to be called selectively (refer to Sec. 7-4).

Another specialized version of a PLL is the well-known stereo decoder.[23] Here a PLL locks onto a 19-kHz pilot tone, which is used to generate a 38-kHz auxiliary carrier for the demodulation of the $L - R$ (left minus right) channel.

The circuits listed in Table 6-1 operate at very different frequency ranges. Typical frequency ranges of the various semiconductor families are shown in Table 6-2. These figures are by no means limiting values, since more and more the frequency spectra of families such as CMOS are being broadened by improved process technologies. Linear PLL ICs can work at frequencies of roughly 35 MHz. Digital PLLs without prescalers can operate at up to about 250 MHz. If prescaler ICs are used, the frequency range of PPLs is extended beyond 1 GHz. There are very fast

**Table 6-2. Frequency Range of PLL ICs\***

| PLL type | Logic family | Frequency range |
|----------|--------------|-----------------|
| Linear | PMOS | 0–5 kHz |
| | Bipolar | 0–35 MHz |
| Digital | CMOS, NMOS | 0 – 5 MHz |
| | TTL | 0–25 MHz |
| | LS-TTL | 0–50 MHz |
| | S-TTL | 0–85 MHz |
| | ECL | 0–250 MHz |
| | ECL + discrete VCO | 0–ca. 1 GHz |
| | ECL + discrete VCO + discrete mixer | >1 GHz |

\*State of the art for 1983.

prescalers in ECL technology operating at frequencies of more than 1 GHz. The fastest integrated VCOs, however, are usable at frequencies up to 250 MHz. PLLs operating at gigahertz frequencies can be obtained by applying VCOs built from discrete components. Operation at even higher frequencies is obtained by using mixing techniques. This will be discussed in more detail in Sec. 7-7.

As will be seen in the following section, the PLL ICs listed in Table 6-1 offer the options of adding external elements or of controlling by external signals.

## 6-2 ADDING EXTERNAL COMPONENTS TO THE INTEGRATED CIRCUITS

In this section, we are given a number of practical hints on how to use the particular features of the various PLL ICs for maximum benefit.

### DC Biasing

The available PLL ICs differ considerably in the methods of generating the dc bias voltages of various pins, such as reference input or VCO output pins. Many ICs are generating these bias levels internally, so the designer has nothing to worry about. With other circuits, however, the user sometimes has to add quite a number of external components just to generate such biasing levels or to block ac signals from "exotic" dc levels. In many cases, only the application notes will show how large the external parts count will be.

The dc levels of an IC's input and output terminals will be worth investigating. For many PLL IC's these bias levels are chosen so that TTL, CMOS, or ECL signals can be connected directly. There are cases, however, where the bias is at such a level that the customer is forced to use capacitive coupling, level shifters, or other tricky devices.

Many IC data sheets claim TTL or CMOS compatibility. The user should check whether this specification applies to all relevant input and output signals or whether it is valid for just one single pin.

### Center Frequency of the VCO

The center frequency $\omega_0$ of a PLL has to be determined by one or more external components. The methods vary from one type of IC to another, but in most cases the center frequency depends on an $RC$ product.

For many PLL ICs the center frequency is given by an external resistor $R_{ext}$ and an external capacitor $C_{ext}$. In many applications, a current source can be used instead of a resistor. Some PLL ICs already have a resistor integrated on the chip, so one capacitor is sufficient to adjust the center frequency. Because integrated resistors (diffused resistors) can be manufactured with only limited accuracy, the corresponding PLL ICs offer a fine-tuning input pin for trimming the center frequency. In most cases a current has to be fed to this control pin.

### Range Switching of the Center Frequency $f_0$

Many PLL ICs offer the possibility of switching the VCO frequency by a control signal applied to pins which are typically called *range select pin* or *binary keying input*. In the simplest case, the frequency of the VCO is switched by a logic signal from some frequency $f_1$ to another frequency $f_2$. Other ICs offer the selection of four different frequencies by two binary control signals. Figure 6-2 shows a portion of a particular PLL IC, where the center frequency $\omega_0$ is determined by one external capacitor, four external resistors $R_1$–$R_4$, and two binary control signals. According to the truth table in Fig. 6-2 the circuit can operate in four different states.

State 1: $A = 0$ and $B = 0$. Only switch $S_1$ is ON; the center frequency is determined by $R_1$ only and has the value $f_1$.

State 2: $A = 0$ and $B = 1$. Switches $S_1$ and $S_2$ are ON. The center frequency is determined by the parallel connection of $R_1$ and $R_2$, where $R_2$ is generally much greater than $R_1$. The center frequency here has the value of $f_1 + \Delta f_2$.

146

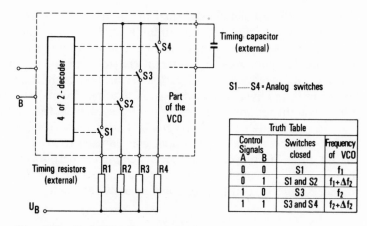

**Fig. 6-2   Switching area of a VCO.**

State 3:  $A = 1$ and $B = 0$. Switch $S_3$ is ON, all other switches are OFF. The center frequency is given by resistor $R_3$ and has a value of $f_2$.

State 4:  $A = 1$ and $B = 1$. Switches $S_3$ and $S_4$ are ON. The center frequency is given by the parallel connection of $R_3$ and $R_4$, where $R_4 \gg R_3$ is most usual. The VCO operates at a frequency $f_2 + \Delta f_2$.

The circuit in Fig. 6-2 can be used as an FSK modulator or demodulator which is able to operate in two different frequency ranges. The frequency range is switched by signal $A$; the $B$ signal is the modulating signal.

### Sweep Inputs for the Center Frequency $f_0$

Some PLL ICs have an input pin that permits sweeping the VCO frequency by an external control signal. The sweep range can be as large as 3 or more decades. An application of the sweep technique was discussed in Sec. 3-4-3 (Fig. 3-22). An interesting application of swept PLLs is found in the "searching" circuits used in modern radio and television.

### Methods of Adjusting the Lock Range $\Delta\omega_L$

As explained in Sec. 3-2-2, the lock range of a PLL is given by the two parameters $\omega_n$ and $\zeta$. These parameters are calculated from the time constants $\tau_1$ and $\tau_2$ of the loop filter and from the VCO gain $K_0$ and the phase-detector gain $K_d$, usually specified in the data sheet [Eq. (3-6)]. Many PLL ICs offer the possibility of tuning $K_0$ or $K_d$ by external elements. The pins provided for this purpose are variously specified as tracking range control, VCO gain control, loop gain control, and so on, where tracking range is synonymous with lock range. With a number of ICs the lock range can be adjusted by an external resistor; in other circuits, a controlling current has to be fed into the control pin (or sunk from the control pin).

Some PLL ICs provide means of limiting the lock range (tracking range limiter). An external jumper can be used to activate this kind of circuit. Such limiting circuits typically clamp the output signal of the loop filter, usually by diodes. These techniques are useful in applications where the lock range of a PLL system is purposely set small.

### Output Signals of the VCO

Most VCO circuits deliver a square-wave output signal. The frequency spectrum of a square wave contains an arbitrarily large number of harmonics, however. In telemetry systems, where a number of data channels are transmitted over one single transmission line, there is a danger that one particular receiver may lock onto a harmonic frequency of another information channel. This problem can be solved by different methods. An obvious solution would be to use a square-wave VCO and filter out the unwanted harmonics. If many information channels are used, the cost of this method becomes prohibitive. An alternative is for a PLL IC to synthesize a staircase VCO signal, which is an approximation to a sine wave.

A third variant is for VCO circuits to output a triangular and a square-wave signal at the same time. A sine wave signal can then be obtained by using a triangular-to-sine converter integrated on the chip. This kind of VCO is offered under the name "waveform generator."

Many PLL ICs offer output signals compatible with familiar logic families such as TTL, ECL, CMOS, and so on. Some circuits even have a "level control pin" enabling the selection of a TTL- or ECL-compatible output signal at the customer's choice.

### PLLs Offering A Quadrature Detector

In some PLL ICs an in-lock detector is integrated on the chip. A quadrature signal (i.e., a signal having a phase offset of 90° with respect to the VCO output signal) is easily obtained with most PLL ICs (see Fig. 6-3).[24] In many VCO circuits a triangular signal is available from the frequency-determining capacitor, so the quadrature signal is readily obtained by using an additional comparator.

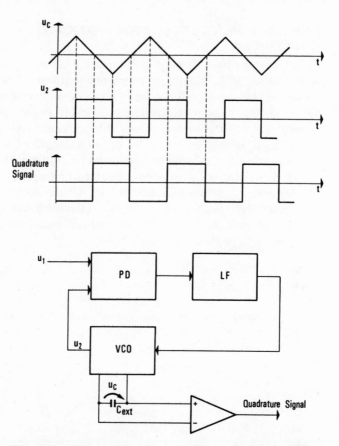

**Fig. 6-3 Generation of a quadrature signal with a supplementary operational amplifier.**

### PLL ICs Having Two Different PDs

Some ICs offer choice between two different kinds of PDs, mostly types 2 and 4 (Table 2-1).

### Additional Functions Integrated on the Chip

Various PLL ICs provide additional operational amplifiers, buffer amplifiers for the output signal of the VCO or of the loop filter, and so on. Some practical applications are given in the next section.

149

## 6-3 PLL INTEGRATED CIRCUITS FOR FREQUENCY SYNTHESIS

As can be seen from Table 6-1, the market offers an impressive number of specialized PLL circuits for frequency synthesis. The operation of frequency synthesizers is discussed in more detail in Sec. 7-7. A number of prescaler ICs are also available, considerably expanding the useful frequency range of the system. There are prescalers having a fixed scaling factor, such as 10. Another kind of prescaler, the so-called 2-modulus prescaler, divides the input frequency by two different factors (such as 10 or 11), depending on an external binary control signal.

An even more complex circuit, the 4-modulus prescaler, is able to scale down the input frequency by four different factors (such as 100, 101, 110, or 111), which are selected by two different binary control signals. Examples will be shown in Sec. 7-7.

# 7 PRACTICAL APPLICATIONS OF THE PLL

**Fully grown-up, curvy, smart
Julia is a real sweetheart!
Daddy boasts: "Now do look here—
See my nubile little dear!"**

Like Busch's Julia, PLL theory has now matured enough to confront real-life problems. The most important PLL applications are discussed in this chapter. Many design examples will be worked out, including oscillograms of typical waveforms or frequency spectra. It is impossible, however, to cover all PLL applications in a book such as this one. Therefore an index of additional PLL literature is included at the end of the text. Let us start our discussion with the tracking filter, which is the heart of numerous PLL applications.

## 7-1 THE TRACKING FILTER

The tracking filter is a bandpass filter whose center frequency is automatically tuned to the frequency of a reference carrier. If we recall the operating principles of the PLL (refer to Fig. 7-1), we note that the PLL does nothing more than *reproduce* a given input signal, where the duplicated signal is out of phase by 90° with respect to the input signal. But what is really the reason for reproducing a signal which already exists? Mr. Pief, an English tourist who came to Germany in the last century thought very much the same way (as told by the author of the story, Wilhelm Busch). When he walked through the country, he exclaimed:

**Why shouldn't I while walking**
**Be distant beauties stalking?**
**It's lovely almost anywhere—**
**What matter if I'm here or there?**

The philosophy did not prove successful, however:

**Blindly he falls into the pool**
**And sees no more, the silly fool!**

Evidently Mr. Pief had not throughly understood the operating principle of the tracking filter! In a similar way, the designer of PLL circuits will better understand the tracking filter after admitting that the input (reference) signal undergoes an effect such as fading (the signal is lost temporarily from time to time). Remember that a PLL behaves like a flywheel—the VCO continues to oscillate at the correct frequency even if the reference is lost temporarily. Unlike Mr. Pief, the PLL can find the right way even if it has lost the ground under its feet. Figure 7-1 shows how the PLL reacts in a situation where the input (reference) signal fades away for a short time.

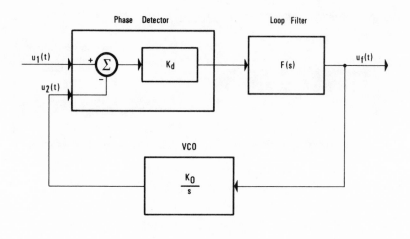

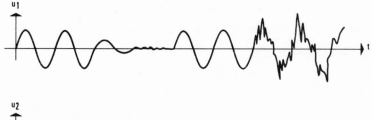

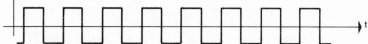

**Fig. 7-1** **Flywheel properties of the PLL. If the reference signal $u_1(t)$ fades away for some time, the VCO continues to oscillate at the last momentary frequency of the reference signal.**

A tracking filter should not only be able to continue to oscillate if the reference signal disappears, it should also be able to remain locked if the input signal is buried in noise. This is explained in more detail with the example of an AM receiver. Figure 7-2 shows a tracking filter which is used in an AM receiver. (The circuit of a completed AM receiver will be discussed in Sec. 7-3.) In this circuit the VCO (Fig. 7-2b) can be tuned over the frequency range of the LW (long-wave) band ($\approx$150–400 kHz) by means of a potentiometer. Figure 7-2a shows the amplified RF signal received by an antenna. Because it is a superposition of a number of information channels, it looks like pure noise. The PLL circuit is now tuned to a frequency where it locks onto one of the carriers in this band. Figure 7-2c shows the waveforms of the VCO output signal (top), a square-wave signal, and the input (reference) signal. The oscilloscope is triggered on the VCO output signal; it can now be clearly observed that the reference

153

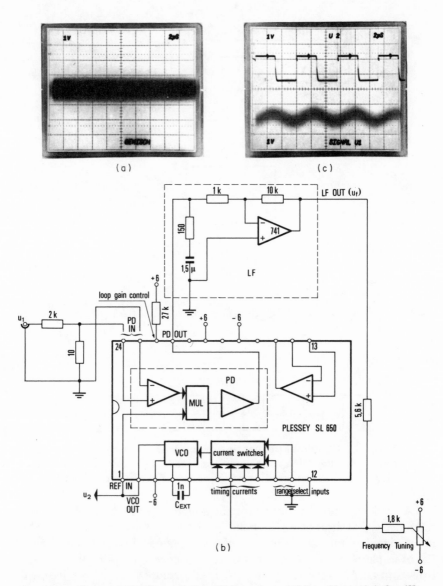

Fig. 7-2 The application of the PLL as a tracking filter. (*a*) Oscillogram of the reference signal $u_1$. (*b*) Practical circuit of a tracking filter. (*c*) Waveforms of VCO output signal $u_2$ (top trace) and reference signal $u_1$ (bottom trace). The oscilloscope is triggered onto the $u_2$ signal; it is clearly visible that the $u_1$ signal has a component that is in phase with $u_2$.

signal contains a frequency component which is synchronous with the VCO frequency.

When considering the circuit of Fig. 7-2b, some unusual features become apparent. The voltage divider (2 kΩ, 10 Ω) seems to be rather unreasonable, but it was needed because the PD in the PLL circuit starts limiting at input levels as low as 4 mV peak. But if this circuit becomes saturated, it is no longer able to suppress noise signals as discussed in Chap. 3. Due to the signal attenuator, the phase-detector gain becomes too small. Therefore the output signal of the PD has to be amplified, which is done by the addition of a type 741 operational amplifier. (The operational amplifier included on the chip has already been used for another purpose not shown in this illustration.)

The loop-filter configuration also looks quite unfamiliar because it consists of just a series connection of $R$ (150 Ω) and $C$ (1.5 μF). Its function becomes evident because the phase-detector output of this IC is a current source, not a voltage source. Furthermore, the external operational amplifier has not been used to build an active loop filter, which would be expected, but is just amplifying the already passively filtered signal. The reason for not utilizing this amplifier as an active filter is its low closed-loop bandwidth. The type 741 would simply not be able to amplify the high-frequency components (in the region of about 100 kHz) of the phase-detector output signal.

The circuit of Fig. 7-2b develops an output signal $u_2(t)$ which is phase-locked to the carrier of one of the transmitters in the frequency band. Hence the square-wave signal $u_2(t)$ can be used for the synchronous demodulation of the audio-frequency (AF) signal; this will be discussed in more detail in Sec. 7-3.

In most tracking-filter applications the signal of particular interest is not $u_2(t)$, however, but rather $u_f(t)$, the output signal of the loop filter. As we know, $u_f(t)$ is the signal that causes the VCO to oscillate at the "correct" frequency, that is, at the frequency of the reference signal. If a PLL is used to demodulate an FM signal, $u_f(t)$ is exactly the AF signal. This is seen easily since an FM reference has the (angular) frequency

$$\omega_1 = \omega_0 + \Delta\omega(t) \tag{7-1}$$

where $\omega_0$ is the carrier frequency and $\Delta\omega(t)$ is the momentary frequency deviation, which represents the AF signal to be transmitted.

According to Eq. (1-1) the frequency of the VCO is given by

$$\omega_2 = \omega_0 + K_0 u_f(t) \tag{7-2}$$

In the locked state $\omega_1 = \omega_2$, and consequently $u_f(t)$ is given by

$$u_f(t) = \frac{\Delta\omega(t)}{K_0} \tag{7-3}$$

i.e., $u_f(t)$ is the *demodulated* signal.

The characteristic of Eq. (7-3) can be easily measured by means of a test circuit as shown in Fig. 7-3. The device under test (DUT) is the very popular PLL IC type CD 4046 first introduced by RCA and manufactured today by many other semiconductor companies. This digital PLL has two different PDs (types 2 and 4) integrated on a chip. For this test the type 2 detector (XOR) was chosen arbitrarily. The system operates at a center frequency $f_0$ = 100 kHz and is laid out for a hold range $\Delta f_H$ = 60 kHz ($\Delta f_H = \Delta \omega_H / 2\pi$) and a lock range $\Delta f_L$ = 20 kHz ($\Delta f_L = \Delta \omega_L / 2\pi$). For the loop filter the simplest $RC$ low-pass filter (type 1 in Table 2-2) is chosen. Following the design rules given in the CD 4046 data sheet (RCA), for the external components we get the following values:

$R_1$ = 22 kΩ

$R_2$ = 100 kΩ

$C_1$ = 390 pF

$R$ = 20 kΩ

$C$ = 1.2 nF

(Refer also to Fig. 7-3.)

The reference signal for the PLL under test (SIG IN) is delivered by a square-wave generator whose operating frequency is swept linearly by a triangular signal $u_{\text{sweep}}$ applied to its sweep input. The output frequency $\omega_1$ of the square-wave generator is then proportional to the sweep signal $u_{\text{sweep}}$. An $XY$ recorder is used to plot the characteristic curve $u_f(\omega_1)$ of the tracking filter. Furthermore the waveforms of the signals $u_1$, $u_2$, $u_d$, and $u_f$ are displayed on a four-channel oscilloscope. The oscilloscope is triggered on the $u_1$ signal.

Now the frequency of the square-wave generator in Fig. 7-3 is swept symmetrically around the center frequency $f_0$ = 100 kHz ($f_0 = \omega_0 / 2\pi$). The sweep rate $\Delta \omega$ was purposely chosen so low that the signals $\overline{u_d}$ and $u_f$ are equal at any time. If the $u_f$ signal is plotted against the reference frequency $f_1$, Fig. 7-4 is obtained. At $f_1 = f_0$ the output signal $\overline{u_d}$ of the phase detector is about half the supply voltage, that is, $V_{DD}/2$ = 5 V ($V_{DD}$ has been chosen to be 10 V). If the reference frequency is increased, the signal $u_f$ rises linearly with $f_1$, as shown by the dashed curve in Fig. 7-4. As soon as $f_1$ exceeds the upper limit of the hold range ($f_0 + \Delta f_H \approx$ 160 kHz), the PLL unlocks, and $u_f$ drops back to about $V_{DD}/2$.

If the reference frequency is reduced, $u_f$ stays at approximately $V_{DD}/2$ (the solid line in Fig. 7-4), but whenever $f_1$ falls below the upper limit of the lock range ($f_0 + \Delta f_L \approx$ 120 kHz), the PLL locks immediately, and the value of $u_f$ jumps to the value given by the dashed line. Similar things happen if the reference frequency is swept in the negative direction, as shown in Fig. 7-4.

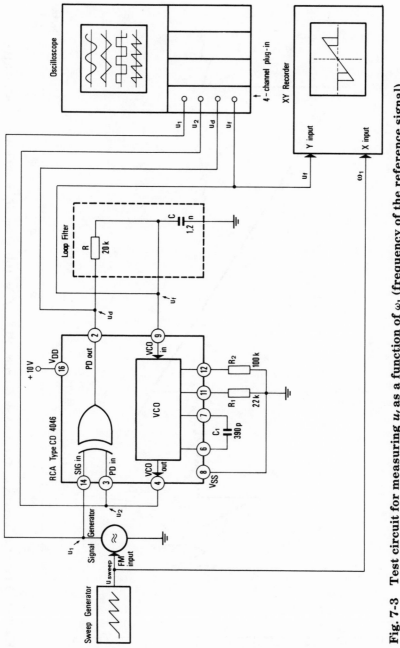

**Fig. 7-3  Test circuit for measuring $u_f$ as a function of $\omega_1$ (frequency of the reference signal).**

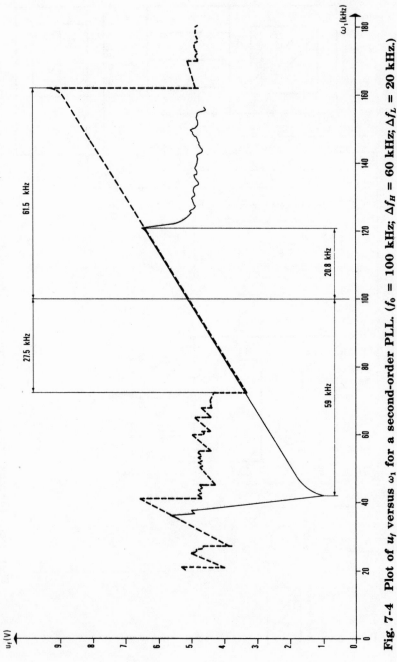

**Fig. 7-4** Plot of $u_f$ versus $\omega_1$ for a second-order PLL. ($f_0 = 100$ kHz; $\Delta f_H = 60$ kHz; $\Delta f_L = 20$ kHz.)

If we examine the dashed line more closely, we notice the appearance of a number of "teeth" in frequency regions where the PLL has already become unlocked. We will come back to this phenomenon very soon.

Let us consider the waveforms measured by the oscilloscope (Fig. 7-5). If the PLL operates at its center frequency $f_0 = 100$ kHz, the signals $u_1$ and $u_2$ are out of phase by 90° (Fig. 7-5a), which corresponds to a phase error $\theta_e = 0$. As expected, the $u_d$ signal is a symmetrical square wave having twice the frequency of the reference signal (200 kHz).

The output signal of the loop filter $u_f(t)$ is simply the "integrated" $u_d$ signal; hence it has a triangular waveform. Its average value $\overline{u_f}$ is approximately $V_{DD}/2 = 5$ V.

Figure 7-5b shows the waveforms for the case where the reference frequency is near the upper limit of the hold range (160 kHz). The signals $u_1$ and $u_2$ are out of phase by nearly 180°, and the phase error is nearly 90°. The signal $u_d(t)$ has a duty-cycle ratio of almost 100 percent. The average value of the loop filter output signal $\overline{u_d}$ is close to the positive supply, approximately 9.5 V.

Figure 7-5c shows the waveforms near the lower limit of the hold range, for a reference frequency of about 40 kHz. By analogy, the duty-cycle ratio of $u_d(t)$ is now near zero percent, and the average value of $\overline{u_f}$ is close to ground potential (approximately 1 V).

A closer look at Fig. 7-4 reveals that there are a number of additional lock ranges besides the "regular" lock range centered around 100 kHz. It shows that the PLL is capable of locking onto harmonics and subharmonics of the center frequency. This phenomenon can be explained by the Fourier spectra of the reference and output signals. If the reference signal is a symmetrical square wave with frequency $f_1$, it contains odd harmonics $3f_1$, $5f_1$, .... If this square wave becomes asymmetrical, even harmonics are also generated. The same holds true for the VCO output signal, which normally produces odd harmonics at frequencies $3f_2$, $5f_2$, .... Consequently the PLL can lock onto subharmonics of the center frequency $f_0$ (refer to Fig. 7-6a.) Here the frequency of the reference signal is one-third of $f_0$, i.e., 33.3 kHz. Its third harmonic (at 100 kHz) is in phase with the VCO output signal and produces a dc term at the output of the PD, as shown by waveforms $u_d$ and $u_f$. The corresponding hold range is clearly visible in the measured curve in Fig. 7-4.

For the same reason the PLL can also lock onto *harmonics* of the center frequency (200 kHz, 300 kHz, etc.). Figure 7-6b shows the waveforms for such a case, but astonishingly the frequency of the reference signal is 350 kHz, which is not a harmonic frequency of $f_0$. What happened here? If we analyzed the $u_2$ signal closely, we would see that the VCO is really oscillating at 100 kHz, but that two subsequent square waves show unequal pulse durations or duty ratios. (These details are not discernible from the rather crude oscillogram of Fig. 7-6.) In fact, the 100-kHz VCO

159

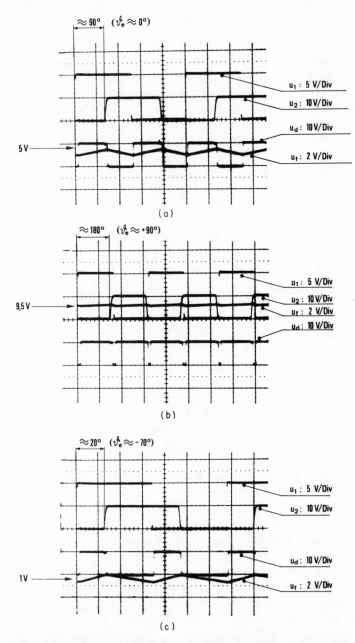

Fig. 7-5 Waveforms of the test circuit of Fig. 7-3. (*a*) The PLL is operating at its center frequency, $f_0 = 100$ kHz. Horizontal scale: 2 $\mu$s/div. (*b*) The PLL is operating near the upper end of the hold range, $f_1 \approx 160$ kHz. Horizontal scale: 2 $\mu$s/div. (*c*) The PLL is operating near the lower limit of the hold range, $f_1 \approx 40$ kHz. Horizontal scale uncalibrated, $> 2$ $\mu$s/div.

frequency has to be considered the second harmonic of a virtual 50-kHz carrier. Consequently, a locking process is always possible when the reference and output frequencies have a common divider (that is, when the ratio of $f_1:f_0$ has the values 1:1, 2:1, 3:1, 1:2, 1:3, 2:3, 3:2, etc. ). A number of additional lock ranges are visible in Fig. 7-4 in the form of additional sawtooth waveforms.

Locking onto harmonics or subharmonics is an unwanted effect, and we must investigate how false locking can be suppressed. The methods depend on the type of PD used. When a type 1 PD is chosen, false locking is always possible if the VCO signal is a square-wave signal. False locking can be inhibited by band-limiting the reference signal. This is not possible, however, if the PLL is to be tunable within a broader frequency band. In this case it would be preferable to use a VCO producing a sine-wave

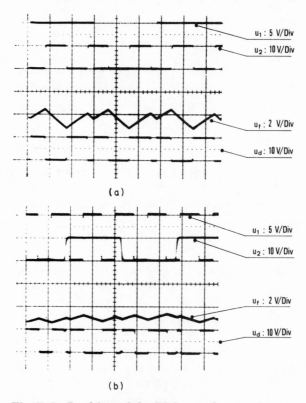

Fig. 7-6 **Locking of the PLL onto harmonics or subharmonics of the center frequency, $f_0 = 100$ kHz. Horizontal scale: 2 $\mu$s/div. (a) Locking onto a subharmonic, $f_1 \approx 33.3$ kHz. (b) Locking onto a harmonic $f_1 = 350$ kHz.**

signal or synthesizing a staircase approximation of a sine wave. These methods are precluded, however, if a type 2 PD is used.

By definition, this PD works only with square-wave signals. If the problem of false locking occurs here, the only reasonable solution would be to replace the type 2 PD with a type 4 PD.

To conclude these considerations, we will analyze the transfer function of the tracking filter. For this purpose it is assumed that the reference carrier $u_1(t)$ is frequency-modulated by a sine-wave signal. The angular reference frequency is then given by

$$\omega_1(t) = \omega_0 + \Delta\omega \sin \omega_m t \tag{7-4}$$

where $\omega_m$ is the modulating frequency and $\Delta\omega$ is the peak frequency deviation. As was seen in Sec. 3-3-2-1, the phase signal $\theta_1(t)$ is obtained by integration,

$$\theta_1(t) = -\frac{\Delta\omega}{\omega_m} \cos \omega_m t \tag{7-5}$$

As long as the PLL operates in its linear region (i.e., for phase errors $\theta_e$ below approximately $\pi/3$), the output signal of the loop filter $u_f(t)$ is also a sine wave having the amplitude $U_f(j\omega_m)$,

$$U_f(j\omega_m) = \frac{j\omega_m H(j\omega_m)\theta_1(j\omega_m)}{K_0} \tag{7-6}$$

where $H(j\omega_m)$ is the phase-transfer function and $\theta_1(j\omega_m)$ is the amplitude of the phase signal.

According to Eq. (7-5), $\theta_1(j\omega_m)$ is given by

$$|\theta_1(j\omega_m)| = \frac{\Delta\omega}{\omega_m} \tag{7-7}$$

For an analysis of the tracking filter we are interested only in knowing the magnitude of the loop-filter output signal. Therefore we obtain

$$\frac{|U_f(j\omega_m)|}{\Delta\omega/K_0} = |H(j\omega_m)| \tag{7-8}$$

This simply means that the gain of the PLL for FM signals is proportional to the phase-transfer function $H(j\omega_m)$ at that modulating frequency.

The gain curve is plotted in Fig. 7-7a for completeness; the modulat-

---

**Fig. 7-7 (Right) Bode plots of a tracking filter. (a) Bode plot of the loop filter output signal $u_f$ versus modulating frequency $\omega_m$. (b) This plot is equivalent to the plot in (a), but here the amplitude of $u_f$ is plotted against *absolute* reference frequency $f_1$ rather than against modulating frequency $\omega_m$.**

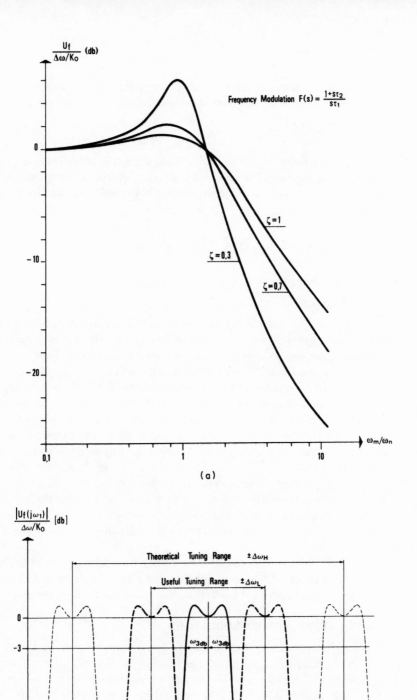

(a)

(b)

163

ing frequency $\omega_m$ is again normalized to the natural frequency $\omega_n$. A type 3 loop filter (Table 2-2) has been specified; its transfer function is

$$F(s) = \frac{1 + s\tau_2}{s\tau_1}$$

Figure 7-7a shows that the PLL acts as a second-order low-pass filter for modulating signals. The gain rolls off at 40 dB per decade above the cutoff frequency $\omega_{3dB}$. The cutoff frequency is by definition the frequency for which the attenuation is 3 dB. It is given by

$$\omega_{3dB} = \left[ 1 + 2\zeta^2 + \sqrt{(1 + 2\zeta^2)^2 + 1} \right]^{1/2} \omega_n \tag{7-9}$$

For optimum damping ($\zeta = 0.7$) $\omega_{3dB}$ becomes

$$\omega_{3dB} = 2.06\omega_n$$

Remember that the modulating frequency $\omega_m$ is superimposed on the carrier frequency $\omega_0$. If we now plot the PLL gain $|H(j\omega_1)|$ as a function of reference frequency $\omega_1 = \omega_0 + \omega_m$, we get the bandpass filter curve of Fig. 7-7b. The 3-dB bandwidth of the bandpass is twice the value of $\omega_{3dB}$.

In contrast to conventional bandpass filters, the PLL is a bandpass with a self-tuned center frequency. Theoretically the tuning range is as large as the hold range $\pm\Delta\omega_H$. It was demonstrated in Sec. 3-2, however, that tracking can be lost if the PLL is operating at the extremes of the hold range. In most practical applications the tuning range is restricted to the lock range $\pm\Delta\omega_L$. The transfer function of the bandpass filter as shown in Fig. 7-7b is easily measured by the test setup of Fig. 7-3. The sweep generator is simply replaced with a sine-wave generator $\omega_m$. The oscilloscope measures the amplitude of the $u_f$ signal vs. the modulating frequency $\omega_m$. A spectrum analyzer will ordinarily be used for this purpose.

The ability of the PLL to self-tune to the center frequency of the reference carrier makes it the ideal solution for FM, PM, or even AM modulators and demodulators. This will be discussed in the next section.

## 7-2 MODULATORS AND DEMODULATORS

A PLL can be operated as a modulator or as a demodulator. It is possible to switch the PLL from modulator operation to demodulator operation, as in modems.[24] The PLL can be used as a communication device for any kind of modulation (AM, FM, PM, or combinations of these) and for any kind of data signals (analog or digital).

There is an enormous quantity of application literature on this topic. Most application notes describe an *arbitrary* method of transmitting information with a PLL, e.g., as by binary FSK or by quaternary PSK (phase shift keying). Normally there is little information indicating which methods of representing and transmitting information are practicable, so the reader is not in a position to justify the feasibility of a particular transmitting scheme. Before entering into details of information transmission with PLLs, let us investigate the various methods of signal representation.

The reason for dealing first with the basic principles of information coding is illustrated by a simple example. A binary signal is to be transmitted over a line in bit-serial form. A logic 1 is given by a HIGH level, a logic 0 by a LOW level. If the information to be transmitted is of the form 10101 ... , the data are easily recognized by the receiver. If on the other hand a data sequence is of the form 100000000001, the data receiver will have some difficulties in deciding how many zeros have actually been sent. There will be further trouble when it comes to differentiating whether a long sequence of LOW levels is either a *real* sequence of logic 0s or just *no information at all.* What is the difference between a 0 and nothing?

A modulated signal can be characterized by a number of parameters:

- Parameter 1 specifies whether the data signal is *analog* or *digital.*

- Parameter 2 specifies the *code* used to represent digital data.

- Parameter 3 describes the *format* of digital data.

- Parameter 4 specifies the *modulation* used.

Parameter 1 differentiates between analog and digital signals. Analog signals are continuously variable, for example in the range of 0–10 V. Digital signals are quantized, however. Binary codes are preferred in most cases, but there are examples of codes having a higher number base, such as ternary and quaternary codes. Very often digital signals are simply quantized representations of analog data (such as measured physical quantities or quantized speech).

Parameter 2 describes the code used to represent digital data. Binary codes are used in most applications. Examples of binary codes are[25]

- Straight binary code (for unsigned integers)

- Sign and magnitude binary code (for signed integers)

- 1s complement binary code

- 2s complement binary code

- Offset binary code

- Gray code

Binary-coded decimal (BCD) codes are frequently used to represent data in decimal form. Alphanumeric data are mostly encoded using the familiar ASCII code, a 7-bit code which can be extended by a parity bit.

165

The code chosen to represent digital data should not be confused with the format, which is characterized by parameter 3.

Parameter 3, the format, specifies the way a particular digital code is physically represented. One of the most popular formats is NRZ (non-return to zero). (Refer to Fig. 7-8.) There are three different NRZ formats

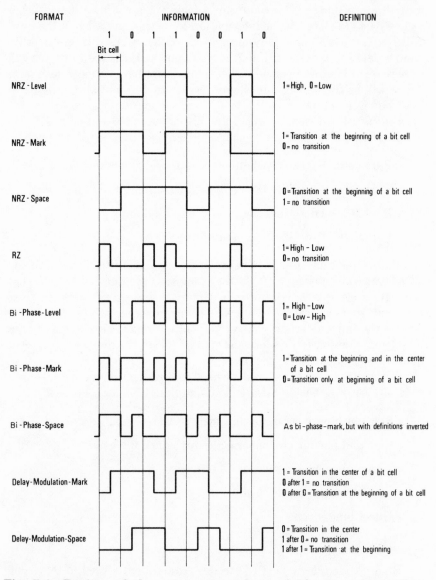

**Fig. 7-8 Review of the most commonly used binary and pseudo-ternary formats.**

in use. With the simplest and most obvious of these, NRZ level, a logic 1 is represented by a HIGH voltage or current level, a logic 0 by a LOW voltage or current level. There are two other NRZ formats: NRZ mark and NRZ space. With the NRZ mark format the signal level is *changed* at the beginning of a bit cell whenever the data to be transmitted is a logic 1. In the case of the NRZ space format, a change of signal level takes place whenever a logic 0 is sent.

The RZ (return-to-zero) format is characterized by the fact that the signal level is always LOW in the second half of the bit cell, as shown in Fig. 7-8. A logic 1 is therefore represented by a HIGH-LOW sequence. For a logic 0 the signal stays LOW throughout the bit cell. In any case the signal level must return to LOW in the middle of the bit cell, hence the name RZ format.

If we compare the RZ and NRZ formats, we observe that with the RZ format positive edges can occur only at the beginning and negative edges only in the middle of a bit cell. We feel that clock signal recovery will be simpler if the RZ format is used. This will be discussed later.

At the receiving end, a message can be correctly detected only if the clock of the transmitted signal is correctly recovered. Clock signal recovery can be a problem with NRZ or RZ formats if a sequence of 0s (or 1s for the NRZ format) is transmitted. The situation can be improved, however, if the message is divided into groups of, for example, 8 bits and a parity bit is added to each group. If odd parity is applied, at least 1 bit in each group of 9 bits is a logic 1. Zero sequences of excessive length are thus avoided, which eases synchronization.

With the NRZ and RZ formats it is not possible to differentiate between logic 0 information and no information at all. To solve the problem, each message is preceded by a unique preamble which tells the receiver that a message will start now. Likewise, messages are generally terminated by a postamble informing the receiver that succeeding LOW levels will in fact represent nothing.

There is a group of other formats called pseudo-ternary, which are actually capable of distinguishing 0 and nothing without preambles. An example is the well-known biphase format shown in Fig. 7-8. Three different biphase formats have been defined. In the most obvious one, the biphase level format, a logic 1 is defined by the sequence HIGH-LOW, a logic 0 by the sequence LOW-HIGH. The other two biphase formats are defined in Fig. 7-8.

For any type of biphase format there is at least one transient in each bit cell, because the signal level always switches in the middle of a bit cell. Hence this format is self-clocking and does away totally with the problem of 0 or 1 sequences. If, furthermore, the signal stays HIGH or LOW for an extended period, the receiver knows that no information at all is transmitted.

The biphase format needs approximately twice the bandwidth of the NRZ code, which is considered a disadvantage. This becomes evident from Fig. 7-8. For an information pattern 10101 . . . , the NRZ format generates a square-wave signal having half the clock frequency, while the biphase level format yields a square-wave signal having the full clock frequency for an information pattern of the form 11111. . . . This disadvantage is overcome by another pseudo-ternary format, the so-called delay-modulation format.[26] The definition of the delay-modulation format is also given in Fig. 7-8. An analysis of the waveforms shows that the required bandwidth of this format is approximately that of NRZ. Moreover there is one transient in each bit cell with the exception of a logic 0 following a logic 1 (for the delay-modulation mark format), hence recovering the clock signal is not a problem.

Figure 7-8 does not list all the possible formats. Besides other binary (or pseudo-ternary) formats, there are a number of higher-valued formats, such as true-ternary, quaternary, octal, hexadecimal, and others. An example of a quaternary format is shown in Fig. 7-9. The four possible states of a quaternary logic signal (0, 1, 2, 3) can be represented by voltage levels of $\pm \frac{1}{2}U$ and $\pm \frac{3}{2}U$, respectively.

The fourth and last parameter of a modulated signal specifies the kind of modulation employed. Let us start with the modulation of analog data signals. Basically there are three methods of modulation: amplitude, frequency, and phase modulation. In the case of amplitude modulation the amplitude of a carrier signal is modulated by the data signal. The unmodulated carrier is given by

$$u_T(t) = \hat{U}_0 \sin \omega_0 t \qquad (7\text{-}10)$$

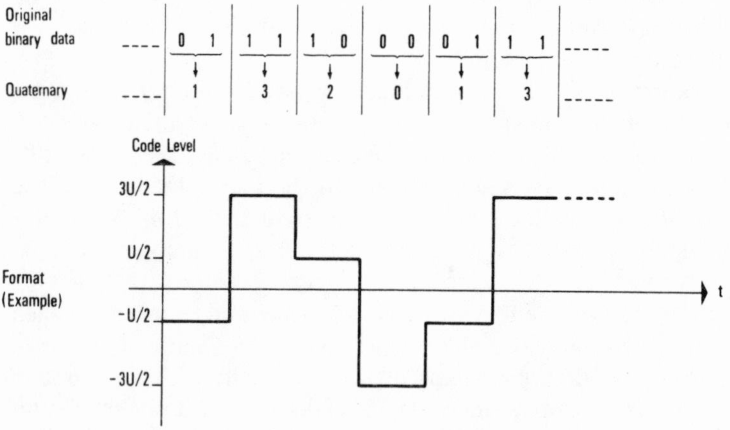

**Fig. 7-9   Example of a quaternary format.**

where $\hat{U}_0$ is the carrier amplitude. The data signal is represented by

$$u_z(t) = \sin(\omega_m t + \phi) \qquad (7\text{-}11)$$

and is normalized for simplicity to the range from $-1$ to $+1$. The AM signal is defined by

$$u_{AM}(t) = u_T[1 + mu_z(t)] \qquad (7\text{-}12)$$

where $m$ is the modulation index.

Substituting Eqs. (7-10) and (7-11) into Eq. (7-12) and making use of the addition theorems of trigonometric functions, we obtain

$$u_{AM}(t) = \hat{U}_0 \left[ \underbrace{\sin \omega_0 t}_{\text{carrier}} + \underbrace{\frac{m}{2} \cos\{(\omega_0 - \omega_m)t + \phi\}}_{\text{lower sideband}} \right. \qquad (7\text{-}13)$$

$$\left. - \underbrace{\frac{m}{2} \cos\{(\omega_0 + \omega_m)t - \phi\}}_{\text{upper sideband}} \right]$$

This is the most general presentation of a double-sideband AM (DSB-AM) with carrier. If the data signal is a sine wave, the frequency spectrum of the modulated signal is a line spectrum according to Eq. (7-13). (See Table 7-1.) It consists of just three lines, the center line representing the carrier, the other two the upper and the lower sidebands. In most cases, however, the data signal is not just a sine wave of fixed frequency, but rather a signal having a broader frequency spectrum which could reach from a lower corner frequency $\omega_1$ to an upper corner frequency $\omega_2$ (Table 7-1). The spectrum of the modulated signal $u_{AM}(t)$ becomes continuous, as shown in the second row in Table 7-1.

Because the upper and lower sidebands of this signal contain the same information, the DSB-AM signal is said to be *redundant*. To recover the data signal, it would be sufficient to transmit only one sideband. The carrier itself is redundant as well, since it does not carry any information. In the case of single-sideband AM (SSB-AM), the carrier and one of the sidebands are filtered away. This is shown in row 4 of Table 7-1, where only the upper sideband is transmitted. The required channel bandwidth is thus reduced by a factor of 2.

To demodulate an SSB-AM signal, the receiver has to reconstruct the carrier. The data signal can then be recovered by mixing down the SSB signal to the baseband, as will be shown in more detail in Sec. 7-3.

To demodulate data signals such as speech, it is sufficient to reconstruct a carrier that has approximately the frequency of the original non-transmitted carrier. If, however, the data signal is a pulse signal, such as a square wave, it can be accurately reconstructed only if the recovered

*Table 7-1 Definition of various modulations, including frequency spectra of the modulated carrier*

| Modulation | Data Signal $u_z(t)$ (LF) — Mathematical representation | Spectrum | Modulated Signal (HF) — Mathematical representation | Spectrum |
|---|---|---|---|---|
| DSB-AM with carrier | $u_z(t) = \sin(\omega_m t + \varphi)$ | | $u_{AM}(t) = u_T\left[\sin\omega_0 t + \frac{m}{2}\cos\left((\omega_0+\omega_m)t+\varphi\right) - \frac{m}{2}\cos\left((\omega_0-\omega_m)t-\varphi\right)\right]$ | |
| DSB-AM with carrier | $u_z(t)$ = random signal (e.g. voice channel) | | | |
| DSB-AM without carrier | $u_z(t) = \sin(\omega_m t + \varphi)$ | | $u_{AM}(t) = \frac{u_T}{2}\left[\cos\left((\omega_0+\omega_m)t+\varphi\right) - \cos\left((\omega_0-\omega_m)t-\varphi\right)\right]$ | |
| SSB-AM | $u_z(t)$ = random signal | | $u_{AM}(t) = u_T\frac{m}{2}\cos\left((\omega_0+\omega_m)t+\varphi\right)$ | |
| VSB-AM | dto. | dto. | | |
| FM | $u_z(t) = \sin(\omega_m t + \varphi)$ | | $u_{FM}(t) = u_0 \sin\left[\left(\omega_0+\Delta\omega\sin(\omega_m t+\varphi)\right)t\right]$ — momentary frequency | |
| PM | $u_z(t) = \sin(\omega_m t + \varphi)$ | | $u_{PM}(t) = u_0 \sin\left[\omega_0 t + \Delta\varphi\sin(\omega_m t+\varphi)\right]$ | |

carrier is exactly in phase with the original carrier. In such a situation it is preferable to employ vestigial-sideband AM (VSB-AM) instead of SSB-AM.

VSB-AM differs from SSB-AM by the fact that the carrier is not completely filtered out and a small portion of the carrier is transmitted. A VSB-AM signal is obtained by first generating a full DSB-AM signal and then filtering away one sideband with a relatively flat filter, as shown by the dashed filter curve in row 5 of Table 7-1. The remaining portion of the carrier can be used to synchronize the reconstructed carrier at the receiver end.

Another kind of amplitude modulation is DSB-AM *with suppressed carrier*. To generate such an AM signal, we simply have to multiply the data signal $u_z(t)$ by the carrier signal $u_T(t)$,

$$u_{AM}(t) = u_z(t)u_T(t) \tag{7-14}$$

Comparing this with Eq. (7-12), we immediately see that no signal having the carrier frequency is generated. If the data signal consists of just one frequency $\omega_m$, the resulting modulated signal will show only the frequencies $\omega_0 + \omega_m$ and $\omega_0 - \omega_m$.

Another important modulation is frequency modulation. We again assume that the data signal is a sine wave given by

$$u_z(t) = \sin(\omega_m t + \phi) \tag{7-15}$$

As in the case of amplitude modulation, its amplitude is normalized to 1. By definition the FM signal is described by

$$u_{FM}(t) = \hat{U}_0 \sin[\{\omega_0 + \Delta\omega u_z(t)\}t] \tag{7-16}$$

where $\Delta\omega$ is the peak (angular) frequency deviation. If Eq. (7-15) is combined with Eq. (7-16), we obtain for the FM signal $u_{FM}(t)$

$$u_{FM}(t) = \hat{U}_0 \sin[\{\omega_0 + \Delta\omega \sin(\omega_m t + \phi)\}t] \tag{7-17}$$

This FM signal is characterized by a constant amplitude. Its momentary frequency, however, is proportional to the momentary value of the data signal.

The FM signal can also be represented by a frequency spectrum. For a sine-wave data signal the FM spectrum no longer consists of just three lines (as was the case with DSB-AM with carrier), but there are (theoretically) an infinite number of spectral lines. The individual frequencies of the spectral lines are $\omega_0 \pm \omega_m$, $\omega_0 \pm 2\omega_m$, ..., $\omega_0 \pm m\omega_m$, ..., where $m$ is an arbitrary positive integer.

The amplitude of the spectral lines decreases with increasing $m$. Mathematically speaking, the amplitude of the $m$th spectral pair of lines is given by

$$A(\omega_0 \pm m\omega_m) = J_m \left( \frac{\Delta\omega}{\omega_m} \right) = J_m(\eta) \tag{7-18}$$

where the function $J_m (\Delta\omega/\omega_m)$ is the Bessel function of order $m$.[27]

The function $J_m(\eta)$ is plotted in Fig. 7-10 for $m = 0, 1, 2$, and 3, and $\eta$ is the modulation index. It allows us to find the relative amplitudes of the first four sidebands for a given modulating frequency $\omega_m$ and a given peak frequency deviation $\Delta\omega$.

We differentiate now between narrow-band and wide-band frequency modulation. For narrow-band frequency modulation the peak frequency deviation is chosen much smaller than the modulating frequency $\omega_m$. Thus $\eta \ll 1$. Only the Bessel functions for $m = 0$ and 1 give essential values. Consequently the frequency spectrum of narrow-band frequency modulation consists roughly of three lines (Table 7-1). The bandwidth of the narrow-band FM signal is therefore approximately

$$B_{\text{FM}} \approx \pm \frac{\omega_m}{2\pi} \quad \text{Hz} \tag{7-19}$$

If $\eta \gg 1$, wide-band frequency modulation is obtained and the frequency spectrum shows many lines.

For a rough approximation it can be said that only those sidebands can be discarded for which $m$ becomes greater than $\eta$. If $\eta = 10$, then 10 pairs of sidebands will be noticeable.

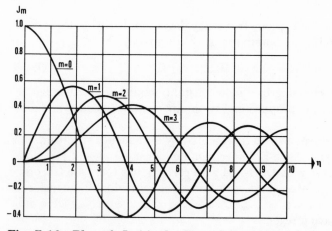

Fig. 7-10  Plot of $J_m (\eta)$, the Bessel functions of the first kind and of order $m$.

The bandwidth of wide-band frequency modulation is approximately

$$B_{\text{FM}} \approx \pm \frac{\Delta\omega + \omega_m}{2\pi} \quad \text{Hz} \tag{7-20}$$

which corresponds approximately to the peak frequency deviation $\Delta\omega/2\pi$.

For phase modulation the modulated signal is written as

$$u_{\text{PM}}(t) = \hat{U}_0 \sin [\omega_0 t + \Delta\Phi u_z(t)] \tag{7-21}$$

where $\Delta\Phi$ is the *peak* phase deviation. According to Eq. (7-21), the momentary phase of the signal is proportional to the data signal $u_z(t)$. If this signal is a sine wave,

$$u_z(t) = \sin (\omega_m t + \phi)$$

the modulated signal becomes

$$u_{\text{PM}}(t) = \hat{U}_0 \sin [\omega_0 t + \Delta\Phi \sin (\omega_m t + \phi)] \tag{7-22}$$

As was true for frequency modulation, the frequency spectrum of $u_{\text{PM}}(t)$ also consists of a number of lines which are spaced by $\omega_m$. (Refer also to Table 7-1.) The amplitude of a particular spectral line at $\omega_0 \pm m\omega_m$ is given by

$$A(\omega_0 \pm m\omega_m) = J_m(\Delta\Phi) \tag{7-23}$$

where $A$ denotes the amplitude.

As for frequency modulation we can state that the number of essential spectral lines increases with the peak phase deviation $\Delta\Phi$. In contrast to frequency modulation, however, the amplitudes of the spectral lines in no way depend on $\omega_m$. This means that the number of spectral lines for a PM signal stays constant, irrespective of the modulating frequency $\omega_m$.

It is easy to see that frequency and phase modulation are so similar that one can be converted into the other. We know that by definition the phase of a signal is the integral of its angular frequency over time $t$,

$$\phi(t) = \int_0^t \omega(t) \, dt$$

(Refer also to Sec 3-1.) Furthermore every VCO is an FM modulator by itself, because its frequency is proportional to the input signal $u_f(t)$. Figure 7-11 demonstrates how an FM modulator is converted into a PM modulator, by simply differentiating the input signal $u_z(t)$.

For $u_f$, therefore, we have

$$u_f = T_d \frac{du_z}{dt}$$

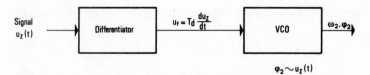

**Fig. 7-11 Interdependence of FM and PM. The block diagram demonstrates how an existing FM modulator can be converted into a PM modulator.**

where $T_d$ is the time constant of the differentiator. The angular frequency of the VCO is consequently

$$\omega_2 = \omega_0 + K_0 T_d \frac{du_z}{dt} \tag{7-24}$$

The phase $\phi_2$ of the VCO is obtained by integration,

$$\phi_2 = \omega_0 t + K_0 T_d \int_0^t \frac{du_z}{dt}\, dt = \omega_0 t + K_0 T_d u_z \tag{7-25}$$

If we set $K_0 T_d = \Delta\Phi$, the output signal of the VCO can be written

$$u_2(t) = \hat{U}_0 \left[ \sin \omega_0 t + \Delta\Phi u_z(t) \right]$$

which is identical to Eq. (7-21).

Thus an FM device is easily converted into a PM device simply by adding a differentiator. By analogy, a PM modulator can be operated as an FM modulator by inserting an integrator.

Let us now discuss the modulation of *binary* data signals. Assume first that a binary signal is used to modulate the amplitude of a carrier signal. If we arbitrarily assign the levels $+1$ and $-1$ to the data signal, corresponding to logic states 1 and 0, respectively, a DSB-AM signal with carrier will have the form

$$u_{AM}(t) = \hat{U}_0 \left( \frac{1 \pm m}{2} \right) \sin \omega_0 t \tag{7-26}$$

Consequently this AM signal will have an amplitude of either $1 + m/2$ or $1 - m/2$. The binary signal contributes the portion $(m/2) \sin \omega_0 t$ or $-(m/2) \sin \omega_0 t$, i.e., a fraction which is either in phase or out of phase with respect to the carrier $\hat{U}_0 \sin \omega_0 t$. It is common practice to represent coherently modulated signals as vectors in a gaussian plane (Table 7-2). The magnitude of such a vector is thereby proportional to the signal amplitude, and its angle with respect to the real axis ($x$ axis) is equal to the phase shift with respect to the carrier. In the case of DSB-AM with carrier, the signal components $(m/2) \sin \omega_0 t$ and $-(m/2) \sin \omega_0 t$ generate

## Table 7-2 Signal state diagrams for a number of different modulations
### Data signal is digital.

| Modulation | Waveform of the modulated Signal | Signal State Diagramm |
|---|---|---|
| DSB·AM₂ with carrier | 1　0　1 | -1 (Data „0")　+1 (Data „1") |
| SSB·AM₂ | 1　0　1 | -1　+1 |
| QAM₄ | in·phase signal　1　0　1 <br> Quadrature signal　0　1　1 | |
| DSB·AM₄ | Signal $+\frac{3U}{2}$ $+\frac{U}{2}$ $-\frac{U}{2}$ $-\frac{3U}{2}$ <br> modulated signal | |
| QAM₁₆ | similar to DSB·AM₄, but with 2 signals having a 90 deg phase difference | |
| FSK₂ | f₁　f₂　f₁　f₂ <br> 1　0　1　1　0 | not applicable |
| PSK₂ | 1　0　1 | -1　+1 |
| PSK₄ | PM·Signal　0　1　3 <br> Reference signal | |
| AM·PM₈ | | |
| AM·PM₉ | | |

175

two vectors with their endpoints at $+1$ and $-1$, respectively, on the $x$ axis (row 1 in Table 7-2). (Note that the scaling on the $x$ axis is arbitrary, so the factor $m/2$ can be dropped.) The vector plot shown in Table 7-2 is usually called a signal state diagram.

In SSB-AM the carrier is suppressed, and a zero phase is not defined. As shown in row 2 of Table 7-2, the modulator signal has constant amplitude. The two signal states are again offset by 180°, but because there is no zero phase defined at the receiving end, their phases are not necessarily 0° and 180°, but can also take arbitrary values of $\phi$ and $\phi + 180°$. Consequently the state diagram of SSB-AM consists of two lines parallel to the $y$ axis, as shown in Table 7-2.

Another very intriguing kind of amplitude modulation is quadrature AM (QAM).[28] Here two carriers are used which are in quadrature (offset by 90°). Both carriers are then independently DSB-AM modulated by two independent binary bit streams. The sum of both modulated signals is finally transmitted, which means that the relative phase of the signal portion can take the values 45°, 135°, 225°, and 315°. At the receiving end both signals are independently demodulated by phase-sensitive detection (see also Sec. 7-3). Note that the usable bandwidth is doubled in the case of QAM. The transmitted QAM signal is effectively a quaternary one, so it has become customary to designate this modulation $QAM_4$.

The channel capacity can also be improved for DSB-AM by admitting more than two levels of the modulating signal. Row 4 of Table 7-2 lists the case of $DSB\text{-}AM_4$. Here the modulating signal is quaternary and shows amplitudes of $+\frac{3}{2}U$, $+\frac{1}{2}U$, $-\frac{1}{2}U$, and $-\frac{3}{2}U$. The signal state diagram consists of four vectors, each being on the $x$ axis. Extending this principle to QAM yields $QAM_{16}$ (row 5 of Table 7-2), namely, amplitude modulation with 16 different signal states. This method is often found in high-speed modems, because it increases the useful bandwidth of cables by a factor of 16.[28]

One of the most popular modulating principles is the well-known FSK. As discussed in Sec. 3-3-2-2, the frequency of an oscillator is shifted between two frequencies $f_1$ and $f_2$ by a modulating signal. A signal state diagram cannot be defined here, because there is no defined phase relationship with reference to a carrier.

The channel capacity can also be increased in the case of FSK by using more than two frequencies. An FSK modulator having four different frequencies, for example, is called $FSK_4$. FSK is very easy to implement by means of PLL techniques, as has already been seen from various examples (see also Sec. 7-4).

The PLL also lends itself readily to the design of phase modulators and demodulators. The simplest PM technique in conjunction with binary data signals is PSK. For $PSK_2$ the phase of a signal is either 0° or 180°

with respect to a reference carrier (see row 7 of Table 7-2). Moreover, at PSK$_4$ the phase of an oscillator is controlled by a quaternary signal and will settle at one of four different values (45°, 135°, 225°, or 315°) with respect to a carrier. The signal state diagrams of PSK$_2$ and PSK$_4$ are also shown in Table 7-2. It can be seen that the signal state diagram of PSK$_4$ is identical with that of QAM$_4$. The potential of PSK can be enhanced even further by using PSK$_8$ or PSK$_{16}$.[28]

It is interesting to note that different modulation schemes can be combined. As an example, amplitude and phase modulation can be utilized at the same time. Table 7-2 lists two possibilities, AM-PM$_a$ and AM-PM$_b$. In both cases the phase has four states, the amplitude two states. In combination, eight states become possible.

Having been confronted with an impressive number of modulation schemes, the user will not find it easy to choose the optimum solution for a given application. The decision would be made easier if some convenient decision criteria could be found.

An important factor is error probability. The error probability tells us how many bits in a bit stream will be corrupted if the stream is transmitted over a given link. If the bit error probability is $10^{-4}$ this means that on the average one in every 10 000 bits will be detected falsely. There are many causes of transmission errors: such as channel-to-channel crosstalk, envelope distortion, reflections, intermodulation, incorrect line equalization, and so on. The sum of all these disturbances can result in either *amplitude distortion* or *phase distortion* of the transmitted signal. These phenomena are plotted in the form of a signal state diagram in Fig. 7-12. The figure illustrates the four signal states found with PSK$_4$ or QAM$_4$ (Table 7-2). Amplitude distortion changes the magnitude of the signal vector, phase distortion alters its phase angle. The endpoints of the signal vectors are thus "smeared" over some area. In order to detect the correct signal, a particular decision zone is assigned to each signal state. In the example of Fig. 7-12 the four decision zones are identical with the four quadrants. A transmission error occurs whenever the endpoint of a vector has migrated—due to noise or distortion—into the decision zone of another signal state. As is easily recognized by considering the signal state diagrams in Table 7-2, the susceptibility to amplitude or phase errors differs greatly for the different modulation schemes. For DSB-AM$_2$ with carrier, a phase error up to $\pm 90°$ can be tolerated, but for QAM$_4$ and PSK$_4$ this error may be no larger than $\pm 45°$.

### SUMMARY OF SEC. 7.2

Analog and digital data signals can be transmitted by means of amplitude, frequency, or phase modulation or by combinations of these. The choice

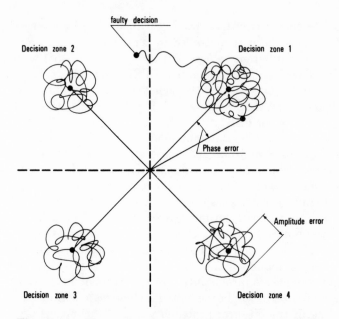

**Fig. 7-12 The influence of amplitude and phase noise on the bit-error rate in digital signal transmission.**

of a particular modulation depends on various criteria, such as the required bandwidth, the bit error probability, and so on.

The functional blocks required to transmit binary information by means of PLL techniques are shown in Fig. 7-13. Either the data come immediately from a digital data source, or they are digitized analog measurements resulting from some kind of sensor.

On the transmitter site, the digital signal is first converted into the desired format. The formatted signal is then modulated. On the receiver site, the incoming signal is first demodulated. To reconstruct the original signal, we must first recover the data clock.

Sections 7-3 to 7-6 will describe the application of the PLL in data communications in more detail.

### 7-3 AM MODULATION AND DEMODULATION

The PLL is inherently a FM modulator or demodulator. Its application to amplitude modulation is less obvious, though easily realized.[29]

The very first AM modulators were built with electron tubes. The familiar transistor modulators such as the base modulator (Fig. 7-14a) and the collector modulator (Fig. 7-14b) emerged from the old tube cir-

178

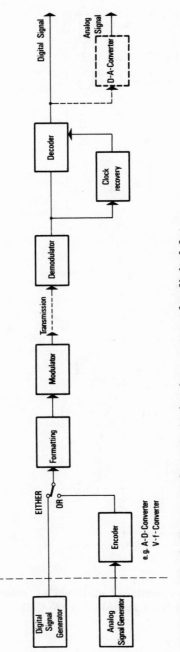

Fig. 7-13  Block diagram of a communications system for digital data.

179

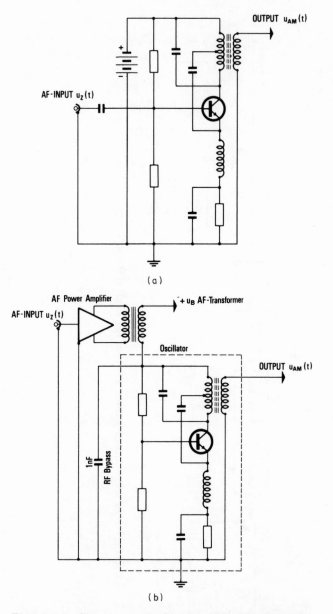

Fig. 7-14  Some conventional methods of amplitude modulation. (*a*) Base modulation. (*b*) Collector modulation.

cuits. In base modulation the nonlinearity of the base-emitter junction is used to modulate the collector current with the data signal. This method is prone to intermodulation distortion, however. A better linearity is obtained with collector modulation. Collector modulation makes use of the fact that the output signal of an oscillator is proportional to its supply voltage. Hence, the data signal is used to deliver the supply voltage of the power oscillator. Consequently the data signal amplifier has to supply the full power of the RF output stage.

The PLL offers a much more elegant solution to building an AM modulator (Fig. 7-15). The analog multiplier shown in the block diagram here is not used as a PD, but serves instead to multiply the carrier signal $u_T(t)$—generated by the VCO—by the data signal $u_z(t)$. [Refer also to Eq. (7-12).].

If the data signal stays on a dc level, the multiplier in Fig. 7-15 generates a DSB-AM signal with a carrier. If the dc level is made zero, however, the modulator generates the sidebands only, and the carrier is suppressed. In this case the multiplier operates like a ring modulator.

As already discussed, most VCOs generate square-wave signals; in communications sine waves are usually required. Figure 7-16 shows a circuit generating an AM sine-wave signal. A special PLL IC is used (XR-2206, manufactured by EXAR); a VCO generates a square wave (sync out). A triangle-sine converter is also included and used to output either a traingular or a sine wave which is offset by 90° from the square-wave output. The user determines by an external resistor whether a triangular wave or a sine wave will be generated (refer to the XR-2206 data sheet). When a sine wave has been selected, the total harmonic distortion can be minimized by the two trimmers $R_A$ and $R_B$. The pin labeled AM-input is used for modulating the VCO input signal. The user is required to establish a bias voltage, delivered by potentiometer $R_D$, and an ac signal, which is capacitively coupled through $C_C$. The waveforms and frequency spectra of the circuit of Fig. 7-16 are displayed in Fig. 7-17.

The waveforms of the square-wave and sine-wave singals are seen in Fig. 7-17a. There is a phase shift of 90° between these signals. The fre-

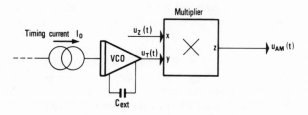

**Fig. 7-15 Application of the PLL as an AM modulator.**

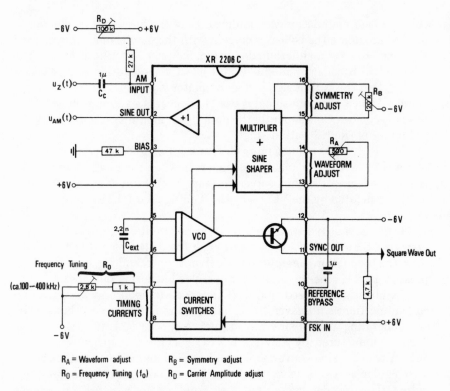

**Fig. 7-16  Practical PLL circuit for AM with carrier or with suppressed carrier.**

quency spectrum of the sine wave is shown in Fig. 7-17*b*. Theoretically the spectrum should contain just one single line, but due to distortion in the sine converter, harmonics (specially the second and third) are clearly visible.

It should be noted that tuning the sine converter for minimum harmonic distortion is accomplished more easily if the spectrum is considered rather than the waveform. A THD (total harmonic distortion) of 5 percent is scarcely observed when we look at the waveform, but it is easily detected when we look at the sidelobes in the frequency spectrum.

---

**Fig. 7-17 (Right)  Waveforms and frequency spectra of the circuit of Fig. 7-15. (*a*) Sine- and square-wave outputs of the VCO (175 kHz). (*b*) Frequency spectrum of the sine wave. Horizontal scale: 200 kHz/div. Note that the marker in the center of the frequency scale is not a spectral line, but is used only to identify the origin of the frequency scale. (*c*) Frequency spectrum of the square-wave output of the VCO. Horizontal scale: 500 kHz/div. (*d*) AM signal, DSB-AM with carrier, $m = 0.3$.**

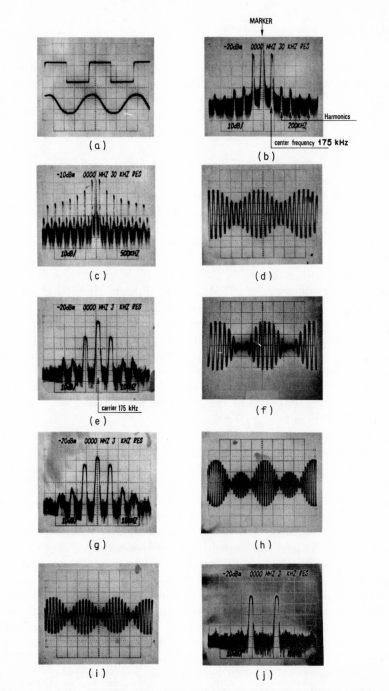

(*e*) Frequency spectrum of the AM signal of (*d*). (*f*) AM signal, DSB-AM with carrier, $m \approx 1$. (*g*) Frequency spectrum of the AM signal of (*f*). (*h*) AM signal, DSB-AM with carrier, $m > 1$. (*i*) AM signal, DSB-AM with suppressed carrier. (*j*) Frequency spectrum of the AM signal of (*i*).

For completeness the frequency spectrum of the square-wave signal is shown in Fig. 7-17c. As expected, it contains mainly odd harmonics. The amplitude of the spectral lines decays with $1/f_m$. Due to the nonideal waveform of the square wave, residual even harmonics are also visible in the spectrum. Note that different frequency scales have been chosen for Fig. 7-17b and c.

Up to now only spectra of unmodulated signals have been considered. Waveforms and spectra of AM signals are illustrated in the following figures. Figure 7-17d shows an AM carrier having a modulation factor of 30 percent ($m \approx 0.3$). The data signal used is also a sine-wave signal. As is to be expected, Fig. 7-17e shows that the frequency spectrum of this signal consists of three lines. The center of the frequency axis is offset by the carrier frequency (175 kHz), as indicated by the marker. The frequency scale on the horizontal axis is 10 kHz per division, and the modulating signal has a frequency $f_m$ of 12 kHz. The modulating circuit does not operate ideally, because higher harmonics are also visible.

If the modulation factor is increased to 100 percent ($m = 1$), the signal shows the waveform of Fig. 7-17f. In this case, the envelope of the modulated signal barely touches the centerline. The corresponding spectrum is given in Fig. 7-17g. The amplitude of the sidebands has increased as compared to the preceding case (Fig. 7-17e). If the modulation factor is raised beyond 100 percent ($m > 1$), the envelopes of the modulated signal cross the centerline (Fig. 7-17h).

The bias potentiometer $R_D$ in Fig. 7-16 is now adjusted to produce zero carrier amplitude, which corresponds to zero dc bias. A DSB-AM signal with suppressed carrier is consequently obtained. Its waveform is illustrated in Fig. 7-17i. Since the carrier amplitude is zero in this case, the modulation factor becomes infinite. The spectrum of this signal is shown in Fig. 7-17j. Note that no carrier can be recognized.

Figure 7-18 is a block diagram of QAM modulation, which was discussed in Sec. 7-2. As mentioned there, two independent data signals $u_{z1}$ and $u_{z2}$ are modulated onto two carriers of the same frequency which are offset by 90°. At the receiving end the two signals are detected by synchronous demodulation, which will be demonstrated later in this section. A VCO generating a sine-wave output signal supplies the first of the two carriers required. The multiplying circuit shown to the left of Fig. 7-18 is used to modulate the first data signal $u_{z1}(t)$. The dc level at its modulating input $x$ is chosen to produce DSB-AM *with* carrier.

A PLL system is used to generate the second carrier $u_T$ /90°. The linear PLL readily serves as a 90° phase shifter, because the reference and output signals are out of phase by exactly 90° if the circuit operates at its center frequency. (Refer also to Sec. 7-7.)

If an active loop filter having a pole at $s = 0$ is used, the phase error

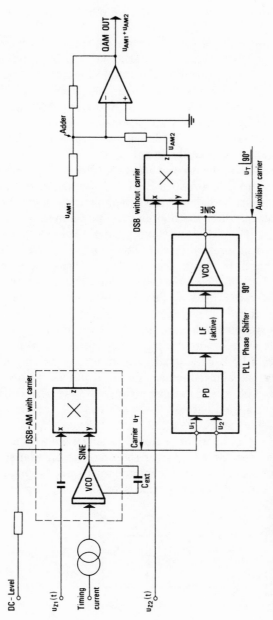

**Fig. 7-18  A QAM modulator built with PLL techniques.**

remains zero, even if the operating frequency is offset from the center frequency. It is therefore convenient to use a linear PLL having an active loop filter for this purpose. Of course, its output signal must be a sine-wave signal.

A second multiplier is used to modulate the auxiliary carrier by the other data signal $u_{z2}(t)$. The dc level at its modulating input is zero here. Thus the carrier is suppressed in the second modulator. But only one of the carriers has to be transmitted, since the receiver should be able to synchronize to one of the carriers. Finally the two AM signals $u_{AM1}$ and $u_{AM2}$ are summed by a summing operational amplifier.

The simple circuit of Fig. 7-18 demonstrates that the bandwidth of a transmission channel can easily be enhanced by a factor of 2.

Similar circuits can be used to generate SSB-AM or VSB-AM signals. The circuits required to produce SSB-AM are generally more complex than those just described. There are two popular methods of generating a SSB-AM signal, the filter method and the phase method. In the case of the filter method (Fig. 7-19) an ordinary DSB-AM modulator is used first, preferably with a suppressed carrier. Then a filter is used to eliminate the unwanted sideband. If VSB is desired, the modulator is designed to produce a carrier. Part of the carrier is then filtered away by the filter.

Until recently, the phase method has been more of academic interest, but its implementation is no longer a serious problem if PLL techniques are applied. A possible implementation is shown in Fig. 7-20. Let the data signal be a sine wave of angular frequency $\omega_m$. A first ring modulator is used to produce a DSB-AM signal with suppressed carrier. (A ring modulator, also called a balanced mixer, is nothing but a multiplier, which was previously discussed, that produces a suppressed-carrier DSB signal.) At the output of ring modulator 1 there are two spectral lines having the frequencies $\omega_0 + \omega_m$ and $\omega_0 - \omega_m$.

According to the theory of ac signals, a sine wave with an angular frequency $\omega$ can be considered a vector rotating with angular speed $\omega$. Consequently the output signal of ring modulator 1 can be represented by two rotating vectors having angular speeds of $\omega_0 + \omega_m$ and $\omega_0 - \omega_m$. The

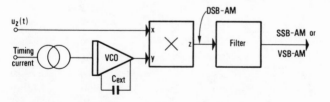

**Fig. 7-19  SSB-AM (or VSB-AM) modulator using the filter method.**

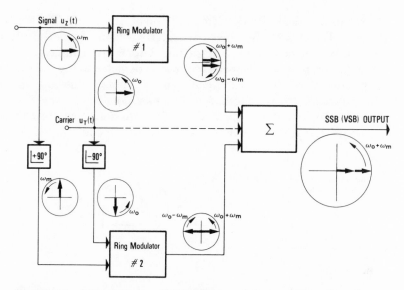

**Fig. 7-20  SSB-AM (or VSB-AM) modulator using the phase method.**

momentary amplitudes of each of the two signals are then given by the projection of the corresponding vectors on the $x$ axis (see Fig. 7-20).

A second ring modulator generates another suppressed-carrier DSB-AM signal. The vector representing its upper-sideband signal is thought for the moment to be in phase with the corresponding signal of ring modulator 1. The vector corresponding to the lower-sideband signal, however, is thought to be in *antiphase*. If this is so, the lower sideband is canceled. The problem is now how to establish the phase relationships as presented in Fig. 7-20. It is solved by shifting the phase of the carrier for the second ring modulator by $+90°$ and the phase of the data signal for this modulator by $-90°$. This is not a difficult task if the signal to be phase-shifted has a constant frequency, but it becomes a problem if an accurate 90° phase shift has to be maintained over a broader frequency band. The realization of such phase-shifting networks is not elementary, though not impossible for an approximation.[30]

Cancellation of the lower sideband is perfect only when the phase shifters are operating without errors. A phase error of only 0.57° causes a spurious sideband amplitude of 1 percent, corresponding to $-40$ dB of upper-to-lower-sideband crosstalk. This is the reason why successful implementation of the phase method still remains problematic. To close the discussion of this circuit, it is easily seen that VSB modulation is obtained by coupling a portion of the carrier to the output of the modulator.

Let us now deal with the demodulation of AM signals. The superiority of the PLL technique becomes particularly evident if one considers first the classical noncoherent AM demodulator. The simplest amplitude demodulator, the so-called peak rectifier, is depicted in Fig. 7-21$a$. We can easily understand that this circuit is not capable of filtering out superimposed noise. Noise cancellation becomes possible, however, if a synchronous demodulator is used (Fig. 7-21$b$). Assume that the carrier frequency of the AM signal is $\omega_0$, and that the modulating signal has a

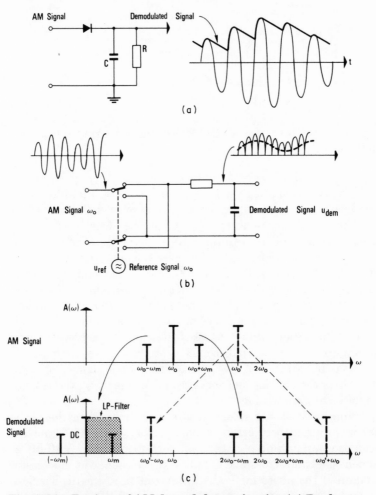

Fig. 7-21 Review of AM demodulator circuits. ($a$) Peak rectifier. ($b$) Synchronous detector (phase-sensitive detector). ($c$) Frequency spectrum of the synchronously demodulated AM signal.

frequency of $\omega_m$. Assume furthermore that a DSB-AM signal with carrier is applied. As illustrated in Fig. 7-21$b$, the polarity of the received AM signal is switched in synchronism with the carrier frequency. As seen from the waveforms, only those signal components that are in phase with the carrier signal can generate a nonzero output signal. This is explained by a simple mathematical analysis. For the DSB-AM signal (with carrier) we can write according to Eq. (7-13)

$$u_{AM}(t) = \hat{U}_0 \left[ \sin \omega_0 t + \frac{m}{2} \cos \{(\omega_0 + \omega_m) t + \phi\} \right.$$

$$\left. - \frac{m}{2} \cos \{(\omega_0 - \omega_m) t - \phi\} \right] \quad (7\text{-}27)$$

For the reference signal, that is, for the reconstructed carrier, we have

$$u_{ref}(t) = \sin \omega_0 t \quad (7\text{-}28)$$

Synchronous demodulation is obtained simply by multiplying the received DSB-AM signal by the reference

$$u_{dem}(t) = u_{AM}(t) u_{ref}(t) \quad (7\text{-}29a)$$

Factoring out this expression yields

$$u_{dem}(t) = \frac{\hat{U}_0}{2} \left[ \underset{\substack{\uparrow \\ \text{dc component}}}{1} + \underset{\substack{\uparrow \\ \text{data signal}}}{m \sin \omega_m t} \right.$$

$$\left. \begin{array}{l} + \dfrac{m}{2} \cos 2\omega_0 t \\[2mm] + \dfrac{m}{2} \sin \{(2\omega_0 + \omega_m) t + \phi\} \\[2mm] + \dfrac{m}{2} \sin \{(2\omega_0 - \omega_m) t - \phi\} \end{array} \right\} \text{ harmonics} \quad (7\text{-}29b)$$

The demodulated signal consists of a dc component, the data signal, and a number of harnomic terms having twice the frequency of the carrier.

The signal spectra in Fig. 7-21$c$ show that synchronous demodulation simply displaces the spectrum by $-\omega_0$ (downward) and by $+\omega_0$ (upward) on the frequency scale. The original data signal is obtained by filtering out the unwanted harmonics by a low-pass filter. Assume now that an interfering signal having a frequency $\omega_0'$ is superimposed on the DSB-AM signal; this disturbance is shown by the dashed spectral line in Fig. 7-21$c$. After synchronous demodulation, this signal appears at the frequencies $\omega_0' - \omega_0$ and $\omega_0' + \omega_0$, as shown in the spectrum.

It becomes evident that those signals that have a spectral line outside the frequency band of interest are *totally suppressed* by synchronous demodulation, provided that the low-pass filter following the synchronous demodulator has a sharp enough cutoff characteristic.

Figure 7-22 shows a synchronous AM demodulator built from a linear PLL circuit. It looks very much like the in-lock detector shown previously in Fig. 3-22. Here the PLL is used to reconstruct the carrier $u_{ref}$ required for synchronous demodulation. Since the phase of $u_{ref}$ will be offset from the original carrier by 90° when the PLL is operating at its center frequency $\omega_0$, a quadrature signal has to be used. A multiplying circuit is used for demodulation. As shown in Chap. 6, such an additional multiplier is included on many PLL chips. The circuit of Fig. 7-22 is an almost complete AM radio receiver for long, medium, and short waves (LW, MW, and SW).

An experimental LW radio receiver (without a power output stage) is shown in Fig. 7-23. This design will also serve as a numerical design example of a linear PLL, including noise considerations. The circuit is designed to operate as a synchronous receiver in the frequency range of 150–400 kHz.

The input signal (RF-IN) of an average radio station is assumed to have an amplitude of approximately 20 mV rms at the RF input of Fig. 7-23. (An RF preamplifier will be required in most cases to provide such a high signal level.) If the receiver has locked onto a particular station, the

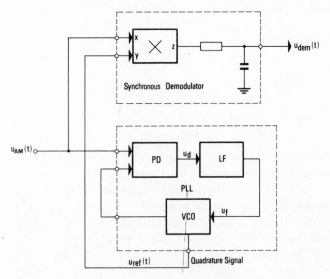

**Fig. 7-22 Synchronous AM demodulator built from a PLL.**

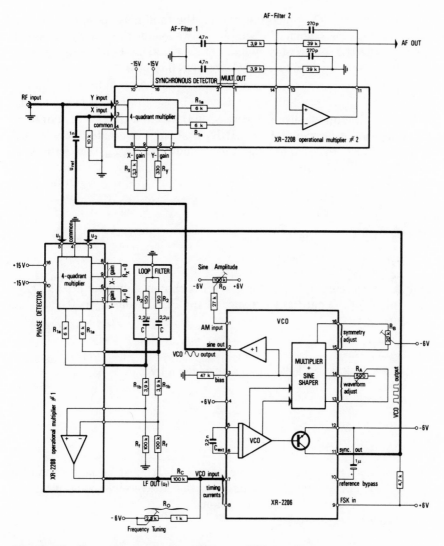

**Fig. 7-23  Experimental circuit for an AM receiver.**

other stations received by the antenna will act on the system as noise sources.

To work out the specifications of the circuit, such as lock range $\Delta\omega_L$, hold range $\Delta\omega_H$, and so on, a reasonable estimate of the SNR at the receiver input should be made. On the basis of some crude measurements the $\text{SNR}_i$ was determined to be approximately

$$\text{SNR}_i = \frac{P_s}{P_n} \approx 0.1$$

[refer to Eq. (3-70)], which means that the total noise power is 10 times that of an average signal. For simplicity, we assuue the noise to be evenly spread over the full frequency band (150–400 kHz). For the signal bandwidth $B_i$, as defined by Eq. (3-68), we therefore obtain

$B_i = 250$ kHz

From simplified noise theory (Sec. 3-4) we know that a PLL can safely lock onto an input signal, if the $\text{SNR}_L$ is about 4 or greater. Now we will use Eq. (3-82) to calculate the noise bandwidth $B_{L \text{ max}}$ that ensures an $\text{SNR}_L$ of 4. We obtain

$$B_{L \text{ max}} = \frac{\text{SNR}_i}{\text{SNR}_L} \cdot \frac{B_i}{2} = \frac{0.2}{4} \frac{250}{2} \approx 15.6 \times 10^3 \text{ s}^{-1}$$

To be on the safe side, we choose

$B_L = 2 \times 10^{-3} \text{ s}^{-1}$

Furthermore, a damping factor $\zeta = 0.7$ is chosen. Applying Eq. (3-77) we can calculate the natural frequency of the PLL. From $B_L = 0.53\omega_n$ (Fig. 3-20) we get

$\omega_n \approx 4 \times 10^{-3} \text{ s}^{-1}$

Using Eq. (3-37), the lock range is now given by

$\Delta\omega_L = 1.4\omega_n = 5.6 \times 10^3 \text{ s}^{-1}$

or

$\Delta f_L \approx 0.9$ kHz

The hold range can now be determined. It must be larger than the lock range, but not excessively large because the receiver could otherwise lock onto the carrier of another station.

$\Delta\omega_H$ was chosen to be

$\Delta\omega_H \approx 84 \times 10^3 \text{ s}^{-1}$

or

$\Delta f_H \approx 13$ kHz

The LW receiver should consequently exhibit the following specifications:

$$f_0 = \frac{\omega_0}{2\pi} = 150\text{–}400 \text{ kHz}$$

$$\omega_n = 4 \times 10^3 \text{ s}^{-1}$$

$$\Delta\omega_H = 84 \times 10^3 \text{ s}^{-1}$$

$$\Delta\omega_L = 5.6 \times 10^3 \text{ s}^{-1}$$

where $f_0$ is the center frequency.

A suitable PLL IC must now be selected. The values of $K_d$ and $K_0$ have to be determined from the data sheets. This enables us to calculate the values of all external components of the circuit of Fig. 7-23. The choice of the PLL IC is crucial. Some earlier experiments had revealed that the multiplier used for synchronous demodulation is the most critical part of the circuit. Nonlinearities of the multiplier give rise to cross modulation, which means that other radio stations will feed through. This problem is aggravated if a square-wave signal is used as the reference $u_{ref}(t)$ for synchronous demodulation. It was therefore decided to use a PLL delivering a sine-wave quadrature signal.

A single-chip PLL IC fulfilling these requirements could not be found, but the circuit can readily be assembled from a few functional block ICs. The intrinsic PLL is built from a VCO IC type XR-2206 (EXAR) and an operational multiplier XR-2208 (EXAR) (Fig. 7-23). The four-quadrant multiplier within the XR-2208 is used as the PD. The loop filter is composed of the internal resistors $R_{1a}$ and the external components $R_2$ and $C$. At first glance this looks like a passive filter. However, it is followed by a difference amplifier comprising the additional operational amplifier and the external resistors $R_{1b}$ and $R_f$. As already mentioned in the discussion of the circuit Fig. 7-16, the XR-2206 simultaneously generates square-wave and sine-wave outputs, out of phase by 90°, which are ideal for this application. The square wave is used as the second input of the PD $(u_2)$, the sine wave for synchronous demodulation.

A second operational multiplier XR-2208 is used for synchronous demodulation. The additional operational amplifier included on the chip is used to amplify the output signal of the demodulator.

The parameters $K_0$, $K_d$, $\tau_1$, and $\tau_2$ should now be determined. The phase-detector gain is given by the gain factor of operational multiplier 1. According to the data sheet, $K_d$ is a function of the gain-setting resistors $R_X$ and $R_Y$. To maximize the gain, $R_X$ and $R_Y$ have been chosen to be zero. Unfortunately the formula for $K_d$ is not valid for this condition. Hence $K_d$ had to be determined experimentally. At a signal level of 20 mV rms, $K_d$ was measured to be 1.5 V rad$^{-1}$. Since the difference amplifier following the passive portion of the loop filter has a voltage gain of 10, the overall phase-detector gain is effectively 15 V rad$^{-1}$. (For the measurement of $K_d$ refer also to Chap. 8.)

This required VCO gain $K_0$ can now be calculated from Eq. (3-28),

$$K_0 = \frac{\Delta\omega_H}{K_d} = \frac{84 \times 10^3}{15} = 5.6 \times 10^3 \text{ s}^{-1}\text{V}^{-1}$$

According to the data sheet, $K_0$ is given by the two external components $R_C$ and $C_{ext}$:

$$K_0 = \frac{2\pi \cdot 0.32}{R_C C_{ext}}$$

Note that the factor $2\pi$ does not appear in the data sheet because the VCO gain is defined here by $K_0 = df_0/du_f$ instead of $d\omega_0/du_f$. If we choose $C_{ext}$ to be 2.2 nF, we obtain an $R_C$ of 100 kΩ, which is well within the recommended range of 1 kΩ to 2 MΩ.

Finally, the two time constants $\tau_1$ and $\tau_2$ have to be calculated. The calculation is made easier if an equivalent circuit of the loop filter is drawn (Fig. 7-24). The actual circuit is depicted at the left (compare also Fig. 7-23), its Thévenin equivalent at the right. The time constant $\tau_1$ is thus given by $(R_{1a}||R_{1b} + R_2)C$, and $\tau_2$ is given by $R_2 C$. Using Eqs. (3-6), we obtain

$$\tau_1 \approx 5 \text{ ms}$$

$$\tau_2 \approx 340 \ \mu\text{s}$$

For $R_2$ and $C$ we get

$$C \approx 2.2 \ \mu\text{F}$$

$$R_2 \approx 150 \ \Omega$$

This completes the design of the PLL circuit. The external components of operational multiplier 2 remain to be determined. The parameters $R_X$ and $R_Y$ are selected so that the circuit never can become saturated;

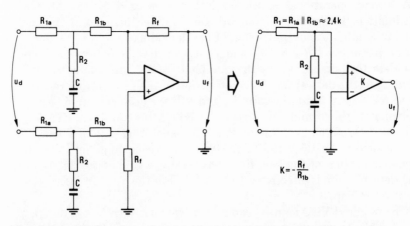

**Fig. 7-24 Derivation of an equivalent network for the loop filter of Fig. 7-23.**

their values are indicated in Fig. 7-23. Furthermore operational multiplier 2 is followed by two filters having a cutoff frequency of 15 kHz.

Some experimental results of the LW receiver are as follows. The PLL locks perfectly onto the various stations when it is tuned by potentiometer $R_0$. The selectivity of the receiver is inferior in comparison with conventional (noncoherent) receivers. The cause is the nonideal performance of the operational multiplier used for synchronous detection. The multiplier exhibits a feedthrough effect from input to output and shows some distortion. These effects lead to cross modulation, which means that channels other than the one selected are weakly coupled to the output. Better results would be obtained if a synchronous filter were used for demodulation (refer to Sec. 7-10).[29]

Figure 7-25 displays the typical waveforms measured by the circuit of Fig. 7-23. The upper traces (Fig. 7-25a) shows the square wave generated by the VCO; the trace in Fig. 7-25b is the output signal of the PD. The latter is deeply buried in noise, but the component contributed by the locked signal is clearly visible. Its theoretical waveform is plotted in Fig. 7-25c. The trace of Fig. 7-25d shows the output signal of the loop filter; if the noise is discarded, a dc level is left.

The trace of Fig. 7-25e shows the input signal $u_1(t)$ when the oscilloscope is triggered by the VCO square-wave output $u_2$. The sine-shaped carrier of the locked radio station is clearly visible.

The next section will deal with the application of the PLL in the domains of FM modulation and demodulation.

## 7-4 FM MODULATION AND DEMODULATION

FM modulation and demodulation are very straightforward applications of the PLL. The block diagram of a generalized FM signal transmission system is depicted in Fig. 7-26. The transmitting portion, shown at the left, is built from the functional blocks of a PLL. The VCO serves as the frequency modulator. The PD, shown as a multiplying element, is here used as a buffer amplifier. Thus, a constant bias level has to be applied to its upper input pin. The receiver is an ordinary linear PLL circuit. There are numerous applications of the circuit shown in Fig. 7-26. The data signal could be delivered by a sensor, such as a temperature or pressure transducer. The transmitter can then be considered a voltage-to-frequency converter. By analogy, the receiver will act as a frequency-to-voltage converter. The overall accuracy of the signal transmission is given by the linearity of the VCOs and their temperature drifts. Because the transmitted signal is an ac signal, it can also be transmitted through transformer-coupled lines, such as telephone lines.

The temperature drift of a VCO is specified by the variation of its

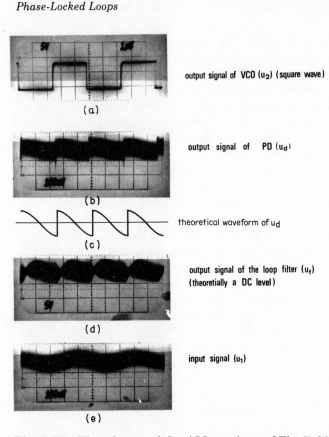

output signal of VCO ($u_2$) (square wave)

(a)

output signal of PD ($u_d$)

(b)

theoretical waveform of $u_d$

(c)

output signal of the loop filter ($u_f$)
(theoretially a DC level)

(d)

input signal ($u_1$)

(e)

**Fig. 7-25   Waveforms of the AM receiver of Fig. 7-23. (*a*)
Square wave output signal of the VCO (*b*) Output signal
of the PD $u_d$. (*c*) Theoretical waveform for $u_d$ assuming
zero noise. (*d*) Output signal of the loop filter $u_f$; (*e*) Input
signal $u_1$. Note that the VCO output signal is repeated in
this oscillogram.**

center frequency with temperature (ppm/°C). The linearity of a VCO is
generally given as the maximum deviation of the frequency-vs.-voltage
characteristic from a straight line, as illustrated in Fig. 7-27.

The linearity of the VCO is rarely specified in data sheets. External
components, such as resistors and capacitors, have an important influence
on the linearity and drift of the system. If precision components are used,
overall error figures of less than 1 percent are easily realized. If considerably higher accuracy is required, the signal should be digitized before
transmission. If only a modest accuracy is needed, the analog system as

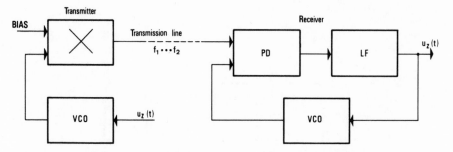

**Fig. 7-26  Communication link using frequency modulation.**

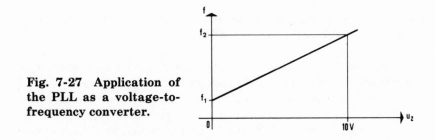

**Fig. 7-27  Application of the PLL as a voltage-to-frequency converter.**

shown in Fig. 7-26 is the simpler solution, because the problem of clock recovery at the receiver end is circumvented.

The data transmission of analog signals—or more precisely, of a carrier modulated directly by an analog signal—can be easily expanded to a number of data channels (Fig. 7-28). The frequency spectrum of the

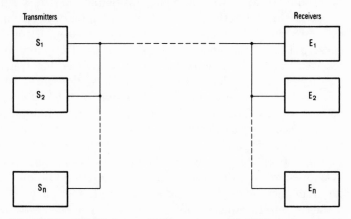

**Fig. 7-28  Multichannel telemetry with PLLs operating as FM transmitters and FM receivers.**

197

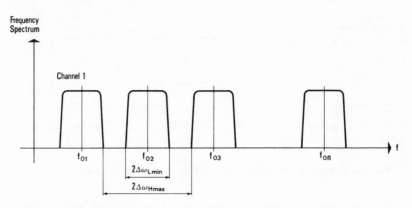

**Fig. 7-29  Frequency spectrum of the multichannel FM telemetry system of Fig. 7-28.**

ensemble of channels is shown in Fig. 7-29. The carrier frequency is in the center of the corresponding data channel. The hold range $\Delta\omega_H$ and lock range $\Delta\omega_L$ must be chosen such that each receiver can lock only onto its particular channel only and not onto adjacent ones.

As indicated in Fig. 7-29, the lock range should be at least as large as the one-sided channel bandwidth, but the hold range should be smaller than the distance from the carrier to cutoff frequency of the adjacent channel. Of course, the stability of each receiver VCO should be high enough to ensure that one channel cannot drift away to another one.

For a particular receiver PLL, the signals of all other channels represent noise. Consequently only linear PLLs can be used as communication receivers. Moreover the $SNR_L$ (as defined in Sec. 3-4) must be good enough to lock onto the desired channel. A numerical example relating to this subject will be presented later in this section. If the data channels cover a broad frequency range, steps must be taken to prevent one receiver from locking onto the harmonics of another transmitter. False locking can be avoided if PLLs are used at the transmitter site whose VCOs are generating sine-wave output signals. If VCOs with square-wave outputs are used, however, the harmonics must be filtered out. Normally it is not necessary to use one filter for each individual channel. As shown in Fig. 7-30, it is possible to use a common filter at the transmitter site for each group of channels whose frequencies are within one octave. This example demonstrates that the harmonics of a group of channels within the frequency band of 1000–1800 Hz are filtered out by a low-pass filter having a cutoff frequency of 2000 Hz. The same can be done for a group of channels within the frequency band of 2000–3800 Hz, and so on.

The configurations shown in Figs. 7-26 and 7-28 are also applicable in cases where the modulating signal is a binary signal. The most obvious modulation scheme for binary data is FSK, as already pointed out. We shall now discuss practical FSK modulator and demodulator circuits.

Two different kinds of FSK demodulators are in general use. The first is an ordinary linear or digital PLL which always stays locked and tracks the frequency of the input signal. The other is a PLL device that locks only when the transmitter is sending one of two frequencies but unlocks when the transmitter switches over to the other frequency. An in-lock detector (as shown in Fig. 6-1) can then be conveniently used as the FSK output stage. This class of FSK demodulators is often referred to as Touch-Tone decoders.

When comparing the properties of both kinds of FSK demodulators, the first is capable of operating at higher speed because it never gets unlocked. The second is slower for it has to establish lock repeatedly, but it can be implemented with fewer external components.

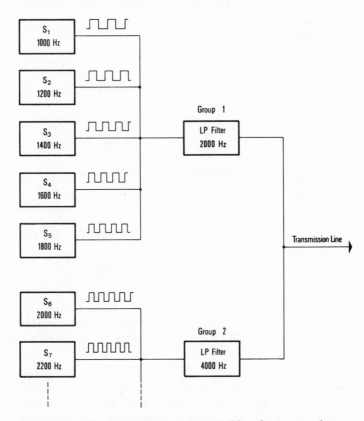

**Fig. 7-30 The harmonics generated by the transmitters have to be suppressed by additional filters.**

There are some very effective Touch-Tone decoder ICs on the market (such as the Computer Microcircuits Ltd. FX-105, refer to Table 6-1). They provide slow data rates but offer excellent noise immunity due to extensive filtering. With this kind of device, multichannel remote-control systems can be realized which operate slowly but are highly reliable and have low bit error rates.

A numerical example of an FSK multichannel telemetry system follows.

### Numerical Example: Multichannel Telemetry with PLLs

A transmission line (telephone line) having a passband from 300 to 3000 Hz is assumed. A number of binary signals, each having a baud rate of 50 baud (1 baud = 1 bit/s), shall be transmitted over the line. All signals are assumed to operate with NRZ format. The following questions are of interest:

1. How many data channels can be transmitted in parallel?

2. How should the individual transmitting and receiving circuits be designed?

### Solution

One of many possible solutions has been actually implemented and is discussed here.

First the required channel bandwidth is determined. For an NRZ format the bit pattern generating the maximum signal frequency is 101010. . . . In this case the signal is a square wave with a frequency of half the bit rate (in our example 25 Hz). The frequency spectrum of the FSK modulation will then display a number of lines spaced by 25 Hz, as discussed in Sec. 7-2.

The signal can be reconstructed at the receiver site if at least the first two sidebands are transmitted. If narrow-band frequency modulation is chosen, the required channel bandwidth is therefore $2 \times 25 = 50$ Hz, and the maximum number of channels is

$$\text{Maximum number of channels} = \frac{3000 - 300}{50} \approx 54$$

This theoretical limit is scarcely realizable. If we worked with the maximum number of channels, there would be no separation between adjacent channels, so it would be very difficult for a receiver to lock onto only one single channel. To obtain better separation, a channel-to-channel spacing of 60 Hz is specified. The number of simultaneous channels is thus reduced to 45.

Assuming that each individual transmitter generates the same output power, the noise power received by any individual receiver is 44 times the signal power of this channel. The SNR at the input of each receiver is then given by

$$\text{SNR}_i = \frac{P_s}{P_n} = \frac{1}{44} \approx 0.023$$

As seen in Sec. 3-4, the $\text{SNR}_L$ should be at least 4 to permit safe locking of the receiver.

The maximum allowable noise bandwidth $B_L$ of the receiving PLLs is calculated from Eq. (3-82),

$$B_L = \frac{\text{SNR}_i}{\text{SNR}_L} \cdot \frac{B_i}{2} = \frac{0.023 \cdot (3000 - 300)}{4 \cdot 2} \text{ Hz} = 7.67 \text{ Hz}$$

For the damping factor the optimum value $\zeta = 0.7$ is chosen. We then obtain the following relationship according to Eq. (3-77):

$$B_L = 0.53\omega_n$$

Hence the natural frequency $\omega_n$ is

$$\omega_n = 14.5 \text{ s}^{-1}$$

or

$$f_n = 2.3 \text{ Hz}$$

For the moment a type 2 loop filter is chosen.

The lock range $\Delta\omega_L$ can now be calculated from Eq. (3-37),

$$\Delta\omega_L = 20.3 \text{ s}^{-1}$$

or

$$\Delta f_L = 3.2 \text{ Hz}$$

Due to the very low $\text{SNR}_i$, the lock range must be made as small as 3.2 Hz. This implies that the long-term drift of the VCO center frequency $f_0$ must be considerably less than 3.2 Hz, that is, in the region of 1 Hz or less. If a greater VCO drift were allowed, the VCO center frequency could run away so far that locking would no longer be possible.

Because we have chosen narrow-band frequency modulation, the peak frequency deviation $\Delta f = \Delta\omega/2\pi$ chosen should be much smaller than the modulating frequency $f_m = \omega_m/2\pi$. A reasonable value is $\Delta f = 5$ Hz. This yields a modulation index $\eta = \Delta f/f_m = 0.2$.

Finally the hold range $\Delta\omega_H$ should be determined. A particular receiver should never lock onto an adjacent channel. Consequently the hold range must be less than 60 Hz,

$$\Delta f_H < 60 \text{ Hz}$$

or

$$\Delta\omega_H < 377 \text{ s}^{-1}$$

An appropriate PLL IC must now be specified. For this example a linear PLL must be selected, such as the XR-215 (EXAR). The schematic diagram of one of the transmitters is shown in Fig. 7-31.

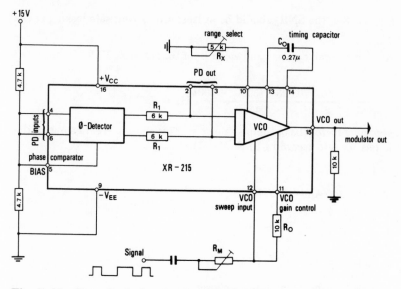

**Fig. 7-31  Practical circuit of an FSK modulator using a PLL IC.**

According to the data sheet of the IC, the center frequency is determined by the external components $C_0$ and $R_X$. $R_X$ is a trimmable resistor in this example. The modulating signal is capacitively coupled to the VCO sweep input. The peak frequency deviation is trimmed by potentiometer $R_M$. The phase-detector and loop-filter sections of the IC are not used at the transmitter site, as shown in Fig. 7-31.

The corresponding receiver circuit is shown in Fig. 7-32. The external components $R_X$ and $C_0$ are identical with those of the transmitter. To enable the suppression of the unwanted channels by the PD, this device is required to operate in its linear region. The output signal level of the transmitters (approximately 2.5 V peak-to-peak) was so large that the phase detectors at the receiving end were operating in saturation. Therefore the signal had to be attenuated at the input of the receivers. Figure 7-33 shows the phase-detector gain vs. input signal level. If the input signal is attenuated to 3 mV peak-to-peak, the phase-detector gain $K_d$ becomes 0.2 V.

Based on this value of $K_d$, the VCO gain of the receiving PLL can now be chosen. According to the data sheet, $K_0$ is determined by the external components $R_0$ and $C_0$ (see also Fig. 7-32),

$$K_0 = \frac{700}{C_0 R_0}$$

where $C_0$ is in microfarads and $R_0$ in kilohms.

The data sheet recommends a value of $R_0$ in the range of 1 to 10 k$\Omega$. $C_0$ is already given (0.27 $\mu$F). Choosing $R_0 = 10$ k$\Omega$ yields a VCO gain of $K_0 = 260$ s$^{-1}$ V$^{-1}$.

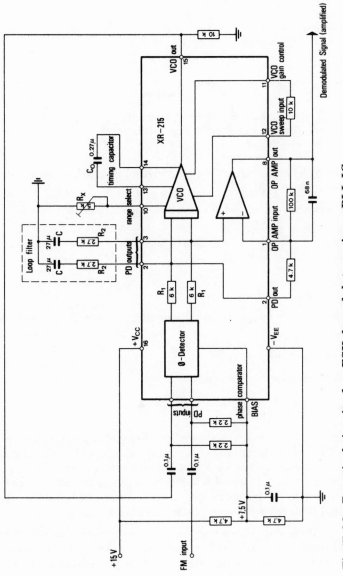

Fig. 7-32 Practical circuit of an FSK demodulator using a PLL IC.

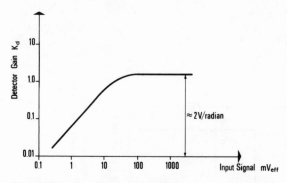

**Fig. 7-33   Detector gain $K_d$ against input signal amplitude for the PD used in Fig. 7-32.**

Consequently the hold range will be $\Delta\omega_H = K_0 K_d = 52$ s$^{-1}$, which is well below the maximum value of 377 s$^{-1}$.

The time constants $\tau_1$ and $\tau_2$ of the loop filter are now calculated:

$$\tau_1 + \tau_2 = \frac{K_0 K_d}{\omega_n^2} = \frac{52}{14.5^2} = 0.25 \text{ s}$$

$$\tau_2 = \frac{2\zeta}{\omega_n} - \frac{1}{K_0 K_d} = \frac{1.4}{14.5} - \frac{1}{52} = 0.08 \text{ s}$$

Resistor $R_1$ is already integrated on the chip (6 kΩ). Hence $C$ (loop filter) will be

$$C = \frac{\tau_1}{R_1} = \frac{0.17}{6 \times 10^3} = 28.3 \ \mu\text{F} \rightarrow 27 \ \mu\text{F}$$

Resistor $R_2$ (loop filter) is obtained from

$$R_2 = \frac{\tau_2}{C} = \frac{0.08}{28.3 \times 10^{-6}} = 2.82 \text{ kΩ} \rightarrow 2.7 \text{ kΩ}$$

Figure 7-34 shows an experimental setup consisting of two transmitters and two receivers. The transmitter channels are summed by an operational amplifier. The required input signal level of 3 mV rms is adjusted by potentiometer $R_F$ at the receiving end. The postfilter used is in Fig. 7-35. The performance of the system is best described by the frequency spectra Fig. 7-36, and 25-Hz square-wave signals are used to simulate the data signals of both transmitters.

Figure 7-36a shows the frequency spectrum on the transmission line when only the first channel ($f_0 = 1000$ Hz) is turned on. The vertical scale is linear. The spectrum shows the carrier at 1000 Hz and the two first sidebands at 1000 ± 25 Hz.

Figure 7-36b shows the same frequency spectrum, but the vertical scale is logarithmic. Second and third harmonics are now clearly visible; the double peak at

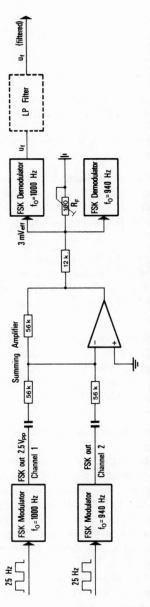

**Fig. 7-34** Experimental multichannel FM telemetry setup using the transmitter and receiver circuits shown in Figs. 7-31 and 7-32. The trimmer $R_F$ serves for the adjustment of the signal amplitude at the input of the receivers.

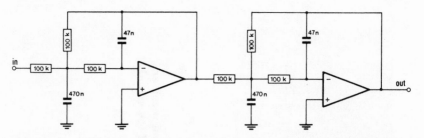

**Fig. 7-35 Postfilter used in the telemetry system of dashed block in Fig. 7-34. A four-pole Chebysheff low-pass filter is used.**

50 Hz probably stems from interference with the (European) line frequency of 50 Hz.

Figure 7-36c again shows the same spectrum, but at a different horizontal scale (500 Hz/div). Since the output signal of the transmitting VCO is a symmetrical square wave, odd harmonics are discernible at 3000 Hz, 5000 Hz, and so on. The even harmonics at 2000 Hz, 4000 Hz, and so on, are much weaker; theoretically they should be zero.

If the second transmitter ($f_0 = 940$ Hz) in Figure 7-34 is also turned on, the spectrum of Fig. 7-36d is obtained. In this display, the vertical scale is linear. Both carriers and two pairs of sidebands are easily recognized. The same spectrum is shown in Fig. 7-36e with the vertical scale logarithmic and the horizontal scale 500 Hz/div.

Now let us look at the spectra at the receiving end. The frequency spectrum of the first receiver—tuned to $f_0 = 1000$ Hz—is recorded. At first only the transmitter with the center frequency of 1000 Hz is switched on. The spectrum of the demodulated signal $u_f$ (Fig. 7-34) is shown in Fig. 7-36f. The 25-Hz peak is clearly visible, together with some higher harmonics.

**Fig. 7-36 Frequency spectra of the multichannel FM communications system of Fig. 7-34. (a) Spectrum of the output signal of the summing amplifier with only transmitter channel 1 switched on. Horizontal scale is 20 Hz/div; vertical scale is linear, 20 mV (rms)/div. (b) Same as (a), but vertical scale is logarithmic, 10 dB/div. (c) Same as (b), but horizontal scale is 500 Hz/div. (d) Spectrum of the output signal of the summing amplifier, both transmitters are on. Horizontal scale is 20 Hz/div; vertical scale is linear, 20 mV (rms)/div. (e) Same as (d), but horizontal scale is 500 Hz/div and vertical scale is logarithmic, 10 dB/div. (f) Spectrum of the demodulated signal $u_f$ in Fig. 7-34 with only transmitter channel 1 switched on. Horizontal scale is 20 Hz/div; vertical scale is 10 dB/div. (g) Same as (f), but with both transmitter channels switched on. (h) Demodulated signal $u_f$ according to (g) after postfilter. Horizontal scale is 20 Hz/div; vertical scale is 10 dB/div.**

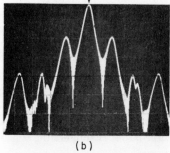

(a)

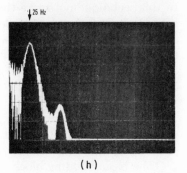

(b)

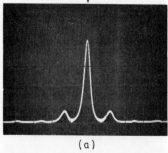

(c)

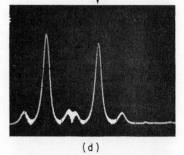

(d)

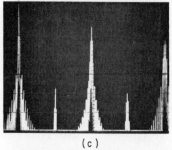

(e)

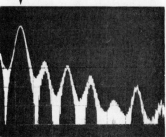

(f)

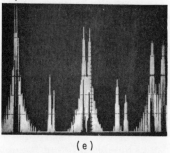

(g)

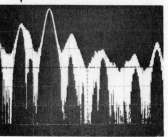

(h)

The second transmitter ($f_0$ = 940 Hz) is then switched on. The resulting frequency spectrum for the $u_f$ signal is shown in Fig. 7-36$g$. The desired peak at 25 Hz is still present, but it is overshadowed by a massive peak at 60 Hz. This 60-Hz peak is generated by the second carrier at 940 Hz. Because the fed-through 60-Hz signal would corrupt the desired 25-Hz signal, it must be filtered out.

One might think of using a loop filter of higher than first order in the circuit of Fig. 7-32, but each additional pole in the PLL's transfer function contributes an additional phase shift of 90° at higher frequencies. To provide stable operation, additional zeros have to be provided which destroy most of the desired steep filter-cutoff characteristic.

It is therefore preferable to locate the filter outside the loop, as sketched by the dashed block in Fig. 7-34. For a practical implementation, a four-pole Chebysheff low-pass filter was built, as shown in Fig. 7-35.[31] Its cutoff frequency is 25 Hz. If the frequency spectrum of the $u_f$ signal in Fig. 7-34 is again measured using the additional filter, the oscillogram of Fig. 7-36$h$ is obtained. The 60-Hz peak is now attenuated by 40 dB and is more than 30 dB below the power level of the 25-Hz signal, which is quite sufficient for digital data signals.

This example demonstrates that a large percentage (typically more than 80 percent of the theoretical bandwidth of the transmission line can be utilized, but that the requirements on the stability of the oscillators can be very demanding (such as ±1-Hz overall stability at $f_0$ = 3000 Hz).

The stability requirements become less stringent if fewer channels are transmitted. The $SNR_i$ becomes larger, so the lock range can also be larger. The solution for the multichannel telemetry system becomes simpler if groups of ten 50-baud channels are combined to form several 500-baud channels. Instead of 45 channels, only 4 or 5 channels must then be transmitted. In this case, drifts which are 10 times greater could be tolerated.

A simpler, but less efficient method of transmitting digital data from multiple sources is given by the so-called *multitone telemetry.* In this technique a particular message (such as "valve off," "over temperature," and so on) is encoded by a sequence of tones having different frequencies. In this technique not more than one transmitter can be active at once, hence the channel capacity is low (typically on the order of 1 baud). Such a low baud rate is nevertheless sufficient for applications such as alarms, monitoring of liquid levels, and so on. Many integrated multitone encoders and decoders are available on the market, such as the FX-407 and FX-507 (Computer Microcircuits Ltd.). These circuits are based on PLL techniques.

There is another class of single- and multitone encoders/decoders which is built by another technique, the zero-crossing technique.[32] The principle of multitone telemetry is explained in Fig. 7-37.

The basic building block of the system is the multitone transceiver,

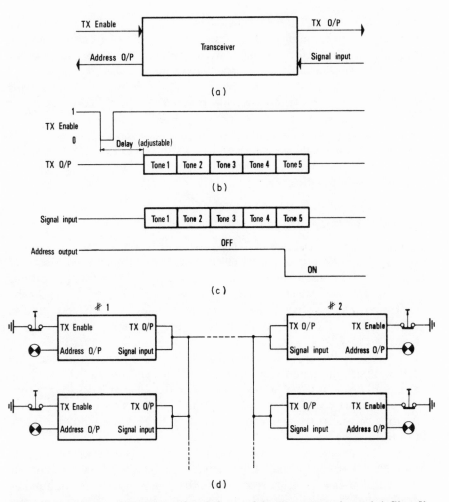

**Fig. 7-37** **Operating principle of the multitone transceiver. (*a*) Simplified block diagram. TX O/P = transmitter output; Signal input = receiver input; TX Enable = start command for the transmitter; Address O/P = status output of the receiver.) (*b*) Operation as a transmitter. (*c*) Operation as a receiver. (*d*) Multichannel telemetry system.**

as shown in Fig. 7-37*a*.[33] This circuit has a transmitter output (TX O/P), a receiving input (Signal input), a triggering input (TX ENABLE), and a status output (Address O/P).

Figure 7-37*b* explains how the transceiver operates in the transmit mode (TX). A trigger pulse applied to TX ENABLE causes a sequence of tones (e.g., five tones) to appear after a selectable delay at the TX O/P. If 10 different tone frequencies are available and if repetitions of the same tone are allowed, $10^5$ different messages can be encoded. Normally each transceiver is programmed to output a unique tone sequence whenever it is triggered. Thus one transceiver would be used to deliver the message "tank full," another for the message "buffer empty," and so on.

Figure 7-37$c$ explains the operation of the circuit in the receiving mode. Whenever the transceiver has detected that tone combination to which it is programmed, it sets Address O/P ON.

An almost unlimited number of transceivers can communicate together via one single transmission line, as shown in Fig. 7-37$d$. Another important application of this type of communication circuit is selective calling in mobile radio systems.

## 7-5 PHASE MODULATION AND DEMODULATION

As already pointed out in Sec. 7-1, phase modulation is related to the frequency modulation by a simple mathematical transformation. Most FM modulator and demodulator circuits can be adopted for PM applications with only minor changes. Consequently, we restrict the discussion to the most important PM technique in data communications, PSK$_2$ (refer also to Table 7-1). The block diagram of a PSK$_2$ communication system is shown in Fig. 7-38.

Parts of a linear PLL are used to build the PSK transmitter. The VCO is operating in an open-loop condition. The multiplier, normally used as PD, is applied as a modulator. The data signal $u_z(t)$ simply controls the polarity of the multiplier output signal in this way. The corresponding waveforms are shown in Fig. 7-39. The PSK demodulator is also built from an ordinary linear PLL.

At the instant when the polarity of the received signal $u_{\text{PSK}}$ is switched, the receiver "sees" a momentary phase error $\theta_e = 180°$. In our mechanical analogy, this corresponds to the case where the pendulum has been "forced" to the upper culmination point (Fig. 3-6). This phase error will be reduced to zero within a short time by the receiver, but it is not known in advance whether the pendulum in the analogy will swing back in the clockwise or counterclockwise direction. Consequently, the output signal of the receiver's loop filter can take the form of either a positive peak or a negative peak when the PLL tracks the 180° phase step. Thus, the polarity of the $u_f$ peak (Fig. 7-39) does not give an indication of whether the data signal has changed from 0 to 1 or from 1 to 0. Additional circuitry is therefore required to reconstruct the data signal (see Fig. 7-38). First the absolute value[34] of the signal $u_f$ is formed. A cascaded Schmitt trigger is used to shape the pulses into rectangular form. The output of the Schmitt trigger finally toggles in the $JK$ flip-flop. Its logic state is changed at every transition of the $u_f$ signal. The original data sig-

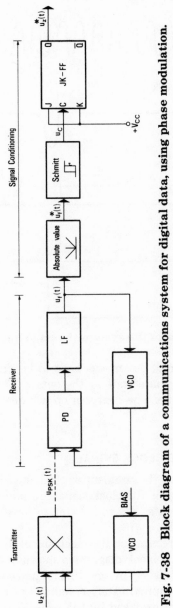

**Fig. 7-38 Block diagram of a communications system for digital data, using phase modulation.**

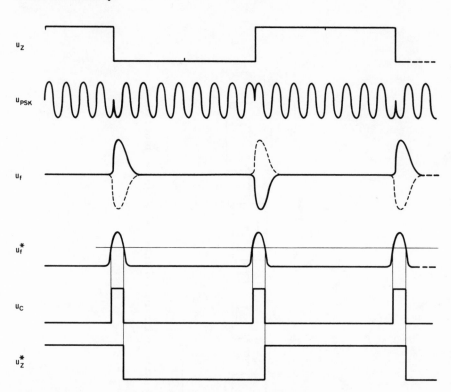

**Fig. 7-39 Waveforms of the system shown in Fig. 7-38.**

nal is thereby recovered, but means have to be provided to set the $JK$ flip-flop to the appropriate state at the outset. This is usually done by some starting procedure. The next section will deal with this problem in more detail.

### 7-6 RECOVERY OF THE CLOCK SIGNAL

This section deals with the problem of clock-signal recovery. The techniques applied to recover the clock signal are largely dependent on the format of the digital data. We will therefore have to consider different clock recovery methods.

Let us start with recovering the clock signal for the RZ format (Fig. 7-8). The corresponding block diagram is shown in Fig. 7-40.[35] Clock-signal recovery is accomplished by one PLL. The center frequency of this PLL is chosen to be approximately equal to the baud rate of the data signal. The PLL is synchronized by the transitions of the demodulated data signal $u_z^*$, as shown in Fig. 7-40. If a type 1 (multiplier) or a type 2 (XOR) PD is used, the VCO generates a square-wave signal which is in

212

quadrature to the demodulated data signal, as seen from the waveforms in Fig. 7-40. This phase relationship is advantageous, for the positive transitions of the VCO output signal can be used to strobe the data signal.

Synchronization of the clock-recovering PLL takes place on every logic 1 contained in the data signal. During a succession of logic 0s the VCO continues to oscillate at its instantaneous frequency. Extended sequences of 0s have to be avoided since the frequency of the VCO could drift away to the extent that synchronization gets lost.

Longer sequences of 0s are avoided if *parity checking* is used. In parity checking an additional bit of information is added to each group of, say, eight successive data bits. When odd parity is chosen, the total number of 1s, including the parity bit, must be *odd*. The choice of odd parity then ensures that in every sequence of nine bits there is *at least one bit* that is not a 0.

One problem still needs attention: the problem of *initialization*. Every message is finished and a pause will follow the message. During the pause no signal is transmitted, a situation which is identical to a long sequence of 0s in the case of the RZ code. When a new message is started, synchronization is likely to be lost. Synchronization must be reestablished; this is done by a fixed preamble which preceeds every message. In the case of the RZ code, a typical preamble consists of a series of 1s.

As stated earlier, the major drawback of the RZ format is the large bandwidth required. The NRZ format needs about half that bandwidth, but clock-signal recovery is slightly more difficult because the frequency

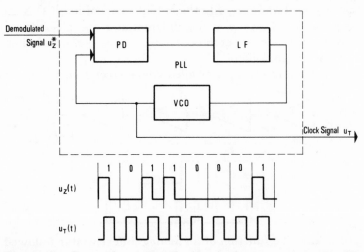

**Fig. 7-40 Operating principle for clock recovery in the RZ format.**

213

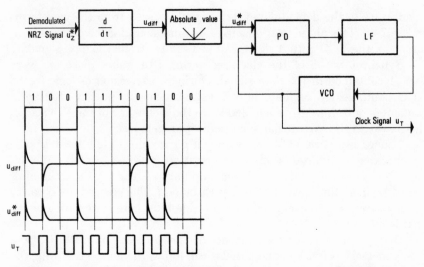

**Fig. 7-41  Operating principle for clock recovery in the NRZ format.**

spectrum of the NRZ does not necessarily contain a spectral line at the clock frequency.[1]

One method of extracting the clock frequency from the NRZ signal is shown in Fig. 7-41. The demodulated NRZ signal $u_z^*$ is first differentiated. Then, the differentiated signal $u_{diff}$ is rectified by an absolute value circuit.[34] As seen in the waveforms in Fig. 7-41, the rectified signal $u_{diff}^*$ contains a frequency component synchronous with the clock. Hence it can be used to synchronize a PLL; the output signal of its VCO is the recovered clock signal $u_T$. Many analog-differentiator and absolute-value circuits have been developed; both operations can be performed alternatively by one single digital circuit.

Figure 7-42 shows a so-called *edge-detector* circuit, in which propa-

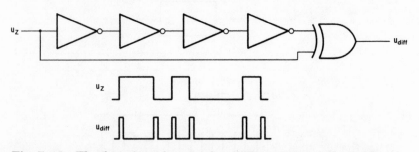

**Fig. 7-42  The function of an analog differentiator followed by an absolute-value circuit can be alternatively performed by a digital edge detector.**

gation delays of gates are used to produce a pulse on each transition of the input signal. The rise and fall times of the input signal must be shorter than the cumulative propagation delay of the four cascaded inverters. If this condition is not met, a Schmitt trigger must be used to clean up the transition.

If the biphase or the delay-modulation format is utilized, the demodulated data signal $u_z^*$ can have transitions at the beginning and in the center of a bit cell, as pointed out in Fig. 7-8. As in the circuits in Figs. 7-40 and 7-41, the demodulated signal $u_z^*$ can be used here to synchronize a PLL operating at twice the clock frequency. A circuit recovering the clock signal for the delay-modulation (DM) format is shown in Fig. 7-43.[26] The waveforms produced by this device are depicted in Fig. 7-44.

An edge-detector circuit produces a short positive pulse $u_{fl}$ on every transition (positive and negative) of the demodulated signal $u_z^*$. The signal $u_{fl}$ is used to synchronize a PLL which operates at twice the clock frequency $2f$ (Fig. 7-44). The output signal of the VCO $(2f)$ is scaled down in frequency by a factor of 2; the rightmost $JK$ flip-flop of Fig. 7-43 is used for this purpose.

The clock signal $u_T$ is now defined to be LOW in the first half of every bit cell and HIGH in the second half. We can see from Fig. 7-43 that the circuit could also settle at the *opposite phase* ($u_T$ being HIGH in the first half of the bit cells). This would result in a faulty interpretation of the received data, because the start and the center or every bit cell are erroneously exchanged.

To establish the correct phase of the recovered clock signal, it is necessary initially to reset the $JK$ flip-flop at the right time. For clarity assume that the recovered clock signal $u_T$ really has the *wrong phase* at the beginning, as shown in Fig. 7-44. An additional circuit is needed, which takes corrective action. Consider again the signal $u_{fl}$ in Fig. 7-44. The time interval between any two consecutive pulses of $u_{fl}$ is not constant, but can show the values of 1, 1½, or 2 periods of a bit cell. An interval longer than 2 periods is impossible. If a three-stage binary UP counter (labeled as the 101 detector in Fig. 7-43) is used to count the (negative) transitions of the signal $2f$ and if this counter is reset by every $u_{fl}$ pulse, its content never exceeds 4. Moreover, as we can clearly see in Fig. 7-44, the content of 4 can only be obtained in the first half of a bit cell, and never in the second half.

This fact is used to properly reset the divide-by-2 flip-flop in Fig. 7-43. Whenever the 3-bit counter reaches 4, a monostable multivibrator (single shot) is triggered. This resets the $JK$ flip-flop. As shown by the waveforms, the phase of the recovered clock is set to its correct state. To enable the corrective action, a preamble of the form 101 ... should precede every message.

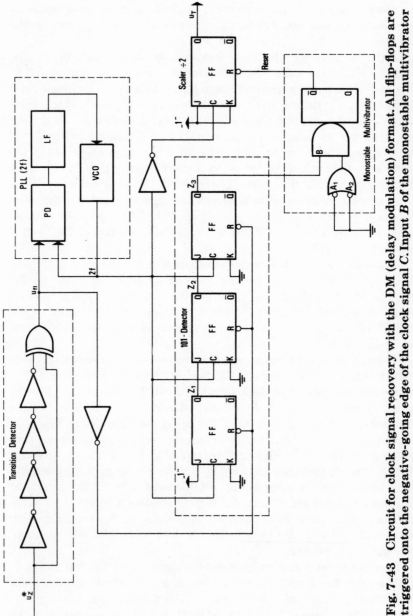

**Fig. 7-43** Circuit for clock signal recovery with the DM (delay modulation) format. All flip-flops are triggered onto the negative-going edge of the clock signal $C$. Input $B$ of the monostable multivibrator triggers on the positive-going edge; inputs $A_1$ and $A_2$ trigger on the negative-going edge. Inputs $A_1$ and $A_2$ are unused here.

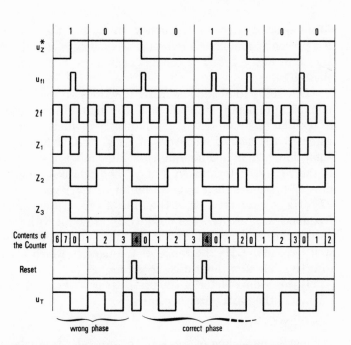

**Fig. 7-44   Waveforms of the circuit of Fig. 7-43.**

It is no major problem to conceive clock-recovering circuits for other formats. As seen from the three examples discussed, the PLL is the key element in each of these applications.

### 7-7 FREQUENCY SYNTHESIS WITH PLLS

Frequency synthesis is one of the major applications of the PLL. In Sec. 4-2 we saw how the digital PLL can be used to generate a frequency that is an integer multiple of a reference frequency. The basic configuration used for frequency synthesis is shown in Fig. 7-45$a$. This system is capable of generating output frequencies which are an *integer* multiple of the reference frequency $f_1$. Later in this section it will be demonstrated that even *fractional* multiples of the reference frequency can be generated.

Frequency synthesizers are found in FM receivers, CB transceivers, television receivers, and the like. In these applications there is a need for generating a great number of frequencies with a narrow spacing of 10, 5, or even 1 kHz. If a channel spacing of 10 kHz is desired, a reference frequency of 10 kHz is normally chosen. Most oscillators should be quartz-crystal-stabilized. A quartz crystal oscillating in the kilohertz region is quite a bulky component. It is therefore more convenient to generate a higher frequency, typically in the region of 5–10 MHz, and to scale it down

217

to the desired reference frequency. In most of the frequency-synthesizer ICs presently available, a reference frequency is integrated on the chip, as shown in Fig. 7-45*b*. The oscillator circuitry is also included on most of these ICs.

One seeks to include as many functions on the chip as possible. It is no major problem to implement all the digital functions on the chip, such as oscillators, phase detectors, frequency dividers, and so on,[36,37] as indicated by the dashed enclosure in Fig. 7-45*b*. Due to its low power consumption, high noise immunity, and large range of supply voltages, CMOS is the preferred technology today. The relatively low speed of CMOS devices precludes their application for *directly* generating frequencies of 27 MHz or more (at least at the time of this writing). To generate higher frequencies, *prescalers* are used; these are built with other IC technologies such as ECL or Schottky TTL (Fig. 7-45*c*). Such prescalers extend to well beyond 1 GHz the range of frequencies which can be synthesized directly, that is, without mixing techniques.[37]

If the scaling factor of the prescaler is $V$ (Fig. 7-45*c*), the output frequency of the synthesizer becomes

$$f_{\text{out}} = NVf_1$$

Obviously the scaling factor $V$ of the prescaler is much greater than 1 in most cases. This implies that it is no longer possible to generate every desired integer multiple of the reference frequency $f_1$; if $V$ is say, 10, only output frequencies of $10f_1$, $20f_1$, $30f_1$, . . . can be generated. This disadvantage can be circumvented by using a so-called dual-modulus prescaler, as shown in Fig. 7-45*d*.[38]

A dual-modulus prescaler is a counter whose division ratio can be switched from one value to another by an external control signal. As an example, the prescaler in Fig. 7-45*d* can divide by a factor of 11 when the applied control signal is HIGH, or by a factor of 10 when the control signal is LOW.

It can be demonstrated that the dual-modulus prescaler makes it possible to generate a number of output frequencies which are spaced only by $f_1$ and not by a multiple of $f_1$. The following conventions are used with respect to Fig. 7-45*d*:

1. Both programmable $\div N_1$ and $\div N_2$ counters are DOWN counters.

2. The output signal of both of these counters is HIGH if the content of the corresponding counters has not yet reached the value 0.

3. When the $\div N_1$ counter has counted down to 0, its output goes LOW and it immediately loads both counters to their preset values $N_1$ and $N_2$, respectively.

4. $N_1$ is always greater than $N_2$.

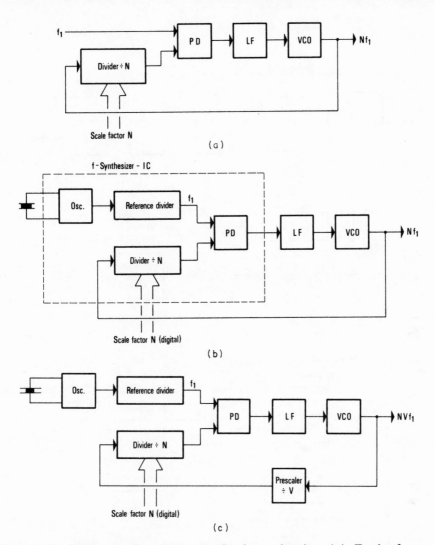

Fig. 7-45 **Different frequency-synthesizer circuits. (a) Basic frequency-synthesizer system. (b) System equal in performance to (a) but with an additional reference divider that makes it possible to use a higher-frequency reference, normally a quartz oscillator. (c) System extends the upper frequency range by using an additional high-speed prescaler. The channel spacing is increased to $Vf_1$.**

5. As shown by the AND gate in Fig. 7-45d, underflow below 0 is inhibited in the case of the $\div N_2$ counter. If this counter has counted down to 0, further counting pulses are inhibited.

The operation of the system shown in Fig. 7-45d becomes clearer if we assume that the $\div N_1$ counter has just counted down to 0 and both counters have just been loaded with their preset values $N_1$ and $N_2$, respectively. We now have to find the number of cycles the VCO must produce

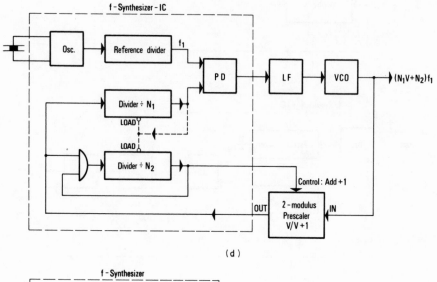

(d)

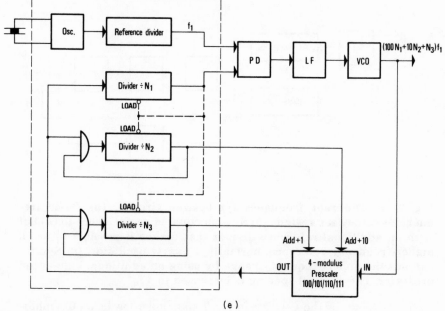

(e)

**Figure 7-45** (*Continued*)

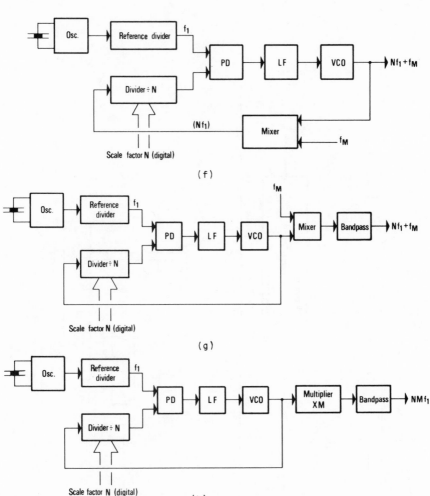

Figure 7-45 (*Continued*) (*d*) System similar to (*c*), employing a dual-modulus prescaler. The channel spacing is reduced to $f_1$. (*e*) System similar to (*d*), but using a four-modulus instead of a dual-modulus prescaler. This not only extends the high-frequency limit, but also allows the generation of lower frequencies. (*f*) System using a mixer in order to expand the high end of the frequency range. (*g*) System similar to (*f*), but with the mixer placed outside of the loop. (*h*) System using an external frequency multiplier for extending the high-frequency end. (*i*) See p. 222. Fractional $N$-loop frequency synthesizer. This system allows the generation of output signals whose frequency is a fractional multiple of the reference frequency $f_{\text{ref}}$. (*j*) See p. 223. Waveforms explaining the operating principle of the fractional $N$ loop.

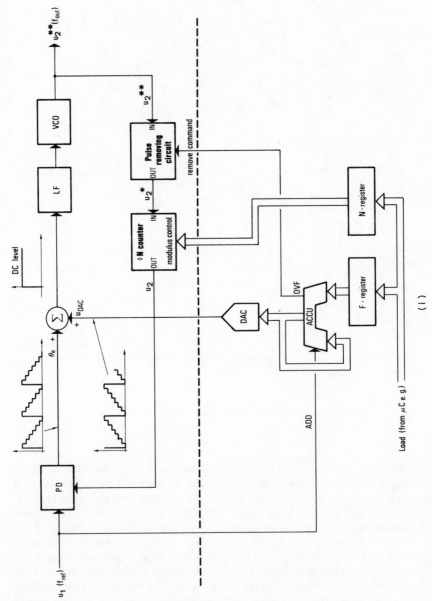

**Figure 7-45** (*Continued*)

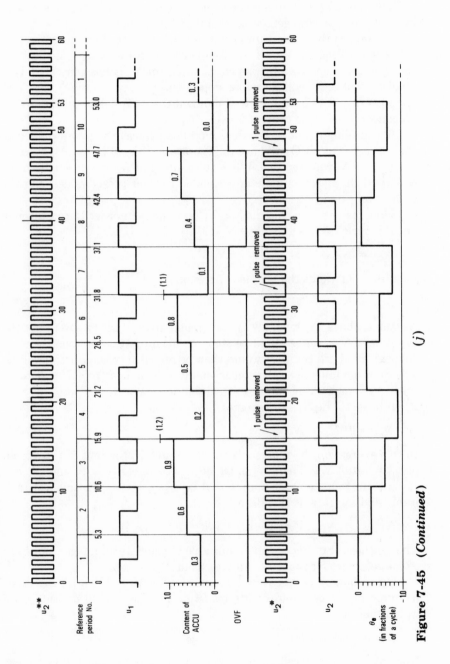

**Figure 7-45** (*Continued*)

223

until the same logic state is reached again. This number is the overall scaling factor of the arrangement shown in Fig. 7-45$d$.

As long as the $\div N_2$ counter has not yet counted down to 0, the prescaler is dividing by $V + 1$. Consequently both the $\div N_1$ and the $\div N_2$ counters will step down by one count when the VCO has generated $V + 1$ pulses. The $\div N_2$ will therefore step down to 0 when the VCO has generated $N_2(V + 1)$ pulses. At that moment the $\div N_1$ has stepped down by $N_2$ counts, that is, its content is $N_1 - N_2$.

The scaling factor of the dual-modulus prescaler is now switched to the value $V$. The VCO will have to generate additional $(N_1 - N_2)V$ pulses until the $\div N_1$ will step to 0. When the content of $N_1$ becomes 0, both the $\div N_1$ and the $\div N_2$ counters are reloaded to their preset values, and the cycle is repeated.

How many pulses $N_{\text{tot}}$ did the VCO produce in order to run through one full cycle? $N_{\text{tot}}$ is given by

$$N_{\text{tot}} = N_2(V + 1) + (N_1 - N_2)V$$

Factoring out yields the simple expression

$$N_{\text{tot}} = N_1 V + N_2 \tag{7-30}$$

As stated above, $N_1$ must always be greater than $N_2$. If this were not the case, the $\div N_1$ counter would be stepped down to 0 *earlier* than the $\div N_2$ counter, and both counters would then be reloaded to their preset values. The dual-modulus prescaler *never* would be switched from $V + 1$ to $V$, so the system could not work in the intended way.

If $V = 10$, Eq. (7-30) becomes

$$N_{\text{tot}} = 10N_1 + N_2 \tag{7-31}$$

In this expression, $N_2$ represents the units and $N_1$ the tens of the overall division ratio $N_{\text{tot}}$. Then $N_2$ must be in the range of 0–9, and $N_1$ can assume any value greater than 9, that is, $N_{1 \text{ min}} = 10$. The smallest realizable division ratio is therefore

$$N_{\text{tot min}} = N_{1 \text{ min}}V = 10 \cdot 10 = 100$$

The synthesizer of Fig. 7-45$d$ is thus able to generate all integer multiples of the reference frequency $f_1$ starting from $N_{\text{tot}} = 100$.

Other factors can of course be chosen for $V$. If $V = 16$, the dual-modulus prescaler would divide by 16 or 17. The overall division ratio would then be

$$N_{\text{tot}} = 16N_1 + N_2$$

Now $N_2$ would be required to have a range of 0–15, and the minimum value of $N_1$ would be $N_{1 \text{ min}} = 16$. In this case, the smallest realizable division ratio $N_{\text{tot min}}$ would be 256.

Let us again assume that $V$ is chosen at 10, and that the (scaled-down) reference frequency $f_1$ of the system in Fig. 7-45$d$ is 10 kHz. The smallest output frequency would then be $100f_1 = 1$ MHz. If the synthesizer IC is built with CMOS technology—the subsystems inside the dashed enclosure are supposed to be CMOS devices—the counting frequencies of the CMOS units will have to be limited to about 3 MHz. Because $V = 10$ has been chosen, no higher frequencies than approximately 30 MHz can be synthesized by the system. This is sufficient for CB radios, but not for digital FM receivers where frequencies higher than 100 MHz have to be generated. [The frequency range of FM radios in 88 to 108 MHz. The frequency synthesizer in a superheterodyne receiver has to generate frequencies which exceed this range by 10.7 MHz (the intermediate frequency), i.e., 98.7–118.7 MHz.]

To extend the frequency range of the synthesizer of Fig. 7-45$d$, we could try to choose a higher value for $V$, say, $V = 100$. The overall division ratio $N_{tot}$ would then be

$$N_{tot} = 100N_1 + N_2$$

where $N_2$ must now cover the range of 0 to 99, and $N_1$ must be at least 100. It should be noted, however, that now the lowest division ratio $N_{tot\ min}$ is no longer 100, but has been increased to

$$N_{tot\ min} = 100N_{1min} = 100 \cdot 100 = 10\ 000$$

If the reference frequency $f_1$ is still 10 kHz, the lowest frequency to be synthesized is now 100 MHz.

Fortunately, there is another way, which extends the upper frequency range of a frequency synthesizer but still allows the synthesis of lower frequencies. The solution is the *four-modulus prescaler* (Fig. 7-45$e$). The four-modulus prescaler is a logical extension of the dual-modulus prescaler. It offers four different scaling factors, and two control signals are required to select one of the four available scaling factors.

As an example, the four-modulus prescaler shown in Fig. 7-45$e$ can divide by factors of 100, 101, 110, and 111.[36] By definition it scales down by 100 when both control inputs are LOW. The internal logic of the four-modulus prescaler is designed so that the scaling factor is increased by 1 when one of the control signals is HIGH, or increased by 10 when the other control signal is HIGH. If both control signals are HIGH, the scaling factor is increased by $1 + 10 = 11$.

As seen from Fig. 7-45$e$, there are no longer two programmable $\div N$ counters in the system, but *three*: $\div N_1$, $\div N_2$, and $\div N_3$ dividers. The overall division ratio $N_{tot}$ of this arrangement is given by

$$N_{tot} = 100N_1 + 10N_2 + N_3$$

In this equation $N_3$ represents the units, $N_2$ the tens, and $N_1$ the hundreds of the division ratio $N_{tot}$. Here $N_2$ and $N_3$ must be in the range 0–9, and $N_1$ must be greater than both $N_2$ and $N_3$ for the reasons explained in the previous example ($N_{1\ min} = 10$).

The smallest realizable division ratio is consequently

$$N_{tot\ min} = 100 \cdot 10 = 1000$$

which is lower by a factor of 10 than in the previous example. For a reference frequency $f_1$ of 10 kHz, the lowest frequency to be synthesized is therefore $1000\ f_1 = 10$ MHz.

Let us examine the operation of the system in Fig. 7-45e by giving a numerical example.

### Numerical Example

We wish to generate a frequency that is 1023 times the reference frequency. The division ratio $N_{tot}$ is then 1023; hence $N_1 = 10$, $N_2 = 2$, and $N_3 = 3$ are chosen. Furthermore, we assume that $N_1$ has just stepped down to 0, so all three counters are now loaded to their preset values. Both outputs of the $\div N_2$ and $\div N_3$ counters are now HIGH, a condition which causes the four-modulus prescaler to divide initially by 111.

### Solution

After $N_2 \cdot 111 = 2 \cdot 111 = 222$ pulses generated by the VCO, the $\div N_2$ counter steps down to 0. Consequently, the prescaler will divide by 110. At this moment the content of the $\div N_3$ counter is $3 - 2 = 1$. After another 110 pulses have been generated by the VCO, the $\div N_3$ counter also steps down to 0. The division ratio of the four-modulus prescaler is now 100.

The content of the $\div N_1$ counter is now 7. After another 700 pulses have been generated by the VCO, the $\div N_1$ also steps down to 0, and the cycle is repeated. To step through an entire cycle, the VCO had to produce a total of

$$N_{tot} = 2 \cdot 111 + 1 \cdot 110 + 7 \cdot 100 = 1023$$

pulses, which is exactly the number desired.

In all frequency synthesizer systems previously considered, multiples of a reference frequency have been generated exclusively by scaling down the VCO output signal by various counter configurations. To produce frequencies in the range of 98.7 to 118.7 MHz with a spacing of 10 kHz, a synthesizer circuit would have had to be designed to offer an overall division ratio of 987 to 1187. As an alternative, one could first generate output frequencies in the range of 8.7 to 18.7 MHz, using a division ratio of 87 to 187, and then *mix up* the obtained frequency band to the desired band. An additional local oscillator operating at a frequency of 90 MHz would be required in this case.

A frequency synthesizer system using an up-mixer is shown in Fig. 7-45f. The basic synthesizer circuit employed here corresponds to the simple system shown in Fig. 7-45b. Of course, all synthesizer systems using dual- and four-modulus prescalers can also be combined with a mixer. In the system of Fig. 7-45*f* the frequency of the local oscillator is $f_M$. Consequently the synthesizer produces output frequencies given by

$$f_{out} = Nf_1 + f_M$$

The mixer is used here to *mix down* these frequencies in the *baseband* $Nf_1$. The mixer also generates a number of further mixing products effectively, generally frequencies given by

$$f_{mix} = \pm nf_{out} \pm mf_M$$

where $n$ and $m$ are arbitrary positive integers.

All these mixing products (excluding the baseband $f_{out} - f_M$) have frequencies that are very much higher than the baseband, so they are filtered easily by either a low-pass filter or even the PLL system itself.

An alternative arrangement using a mixer is shown in Fig. 7-45*g*. In contrast to the previously discussed system, the mixer is not inside but outside the loop. The system generates output frequencies identical with those in Fig. 7-45*f*, but for obvious reasons the desired frequency spectrum has to be filtered out here by a bandpass filter.

Still another way of extending the upper frequency range of frequency synthesizers is given by the *frequency multiplier,* as shown by Fig. 7-45*h*.

Frequency multipliers are normaly built from nonlinear elements which produce harmonics, such as varactor diodes, step-recovery diodes, and similar devices. These elements produce a broad spectrum of harmonics. The desired frequency must therefore be filtered out by a bandpass filter. If $M$ is the frequency-multiplying factor, the output frequency of this synthesizer is

$$f_{out} = MNf_1$$

It should be noted that now the channel spacing is not equal to the reference frequency $f_1$, but to $Mf_1$.

It seems evident that a programmable divide-by-$N$ counter can divide the incoming frequency by an *integer* scaling factor only, and not by a *fractional number,* such as 10.5. Dividing by 10.5 becomes possible, however, if the divide-by-$N$ counter is made to scale down alternately first by 10 and then by 11. On the average this counter effectively divides the input frequency by 10.5. Fractional division ratios of any complexity can be realized (at least theoretically). A ratio of 5.3 is obtained if a $\div N$ counter is forced to divide by 5 in seven cycles of each group of ten cycles and by 6 in the remaining three cycles. A ratio of 27.35 is obtained if a $\div N$ counter is forced to divide by 28 in 35 cycles of each group of 100 cycles and by 27 in the remaining 65 cycles.

A frequency synthesizer capable of generating output frequencies which are a fractional multiple of a reference frequency is called a *frac-*

*tional N loop.*[39] Figure 7-45*i* shows the block diagram of a fractional $N$ loop. Its division ratio is the number $N.F$, where $N$ is the integer part and $F$ is the fractional part. For example, if a division ratio of 26.47 is desired, then $N = 26$ and $F = 0.47$.

The upper portion of Fig. 7-45*i*, separated by a dashed line, shows an ordinary frequency-synthesizer system generating an output signal whose frequency is $N$ times the reference frequency $f_{\text{ref}}$. The block labeled pulse-removing circuit and the summing block $\Sigma$ should be disregarded for the moment. The integer and fractional parts ($N$ and $F$, respectively) of the division ratio $N:F$ are stored in a buffer register as shown in the lower part of Fig. 7-45*i*. The numbers $N$ and $F$ can be loaded via a serial or parallel data link from a microprocessor.

The $F$ register stores a fractional number; $F$ can be stored in any code (binary, hexadecimal, BCD, and so on). If $F$ is represented as a two-digit fractional BCD number, for example, the $F$ register is an 8-bit register, where the individual stages are assigned weights of 0.8, 0.4, 0.2, and 0.1 (tenths register) and 0.08, 0.04, 0.02, and 0.01 (hundredths register).

We will now investigate how a frequency synthesizer can divide its output frequency $f_{\text{out}}$ by fractional numbers. Let us see, for example, how the system can generate an output frequency $f_{\text{out}} = 5.3f_{\text{ref}}$. We can understand the operation of the fractional $N$ loop by examining the waveforms shown in Fig. 7-45*j*. Assume that the output frequency is already 5.3 times the reference frequency, and refer to the waveforms $u_2^{**}$ (VCO output signal) and $u_1$ (reference signal) in Fig. 7-45*j*.

The signal $u_2^{**}$ shows 53 cycles during the time interval when the reference signal $u_1$ is executing 10 reference cycles. (In the following, the term *reference cycle* or *reference period* will be used to designate one full oscillation of the reference signal $u_1$.) Let the system start at the time $t = 0$.

During the time interval where $u_1$ generates its first reference cycle, the $\div N$ counter is required to divide the signal $u_2^{**}$ by a factor of 5.3. This is impossible, of course, so the $\div N$ counter will divide by 5 only. In the first reference period, 0.3 pulse is then "missing." This error (0.3) has to be memorized somewhere in the system; an accumulator (ACCU) is used for this purpose.

The ACCU uses the same code as the $F$ register. If a two-digit fractional BCD format is used, the ACCU is capable of storing fractional BCD numbers within a range of 0.00 to 0.99. As seen from Fig. 7-45*i*, the ACCU adds the fractional number $0.F$ supplied by the $F$ register to its original content whenever the ADD signal performs a positive transition, such as at the beginning of each reference period. If we assume that the initial content of the ACCU was zero at $t = 0$, the ACCU will accumulate an error of 0.3 cycle during the first reference period, indicating that the synthesizer has "missed" 0.3 pulse during the first reference period.

In the second reference period the $\div N$ counter is again required to divide by 5.3. Because this is not possible, it will continue to divide by 5 instead. Since it has already missed 0.3 cycle in the first reference period, the total error has now accumulated to 0.6 cycle. In the third reference period the accumulated error is 0.9 cycle, and in the fourth reference period, it is 1.2 cycles. However, the ACCU cannot store numbers greater than 1; consequently it overflows and generates an OVF signal (Fig. 7.45$i$). The content of the ACCU is now 0.2, as seen in Fig. 7-45$j$. The OVF pulse generated by the ACCU causes the pulse-removing circuit to become active, and the next pulse generated by the VCO is removed from the $\div N$ counter. This pulse removal has the same effect as if the $\div N$ counter divided by 6 instead of 5.

As Fig. 7-45$j$ shows, the ACCU overflows again in the seventh and tenth reference periods. Three pulses will therefore be removed from the $\div N$ counter in a sequence of 10 reference periods. Because the $\div N$ counter divides by 5 forever, $10 \cdot 5 + 3 = 53$ pulses are produced by the VCO during 10 reference periods. This is exactly what was intended. However, one problem has been overlooked. If the VCO oscillates at 5.3 times the reference frequency $f_{ref}$, it produces 5.3 cycles during one reference period. The PD in Fig. 7-45$i$ will consequently measure a phase error of $-0.3$ cycle (or $-0.3 \cdot 2\pi$ rad) after the first reference period in Fig. 7-45$j$. Thus the phase error has *negative polarity* because the reference signal $u_1$ *lags* the signal $u_2$.

After the second reference cycle the phase error has increased to $-0.6$ cycle, and so on. The phase error $\theta_e$ is plotted versus time in Fig. 7-45$j$; its waveform looks like a staircase.

This phase error is applied to the input of the loop filter and will modulate the frequency of the VCO. Such a staircase-shaped modulation of the VCO frequency is not desired, however, because the pulse-removing technique just discussed has already compensated for this phase error. There is an elegant way to avoid this undesired frequency modulation: The waveforms of Fig. 7-45$j$ show that the content of the ACCU has the same amplitude as the phase error $\theta_e$ but opposite polarity. The content of the ACCU is therefore converted to an analog signal by a DAC (digital-to-analog converter); the output signal $u_{DAC}$ is added to the output signal of the phase detector. Since the two staircase signals cancel each other, the input signal to the loop filter is a dc level when the fractional $N$ loop has reached a stable operating point.

## 7-8 USING THE PLL AS A PHASE SHIFTER

The application of the PLL as a phase shifter is not a familiar one. Of course it is easier to fabricate a phase shifter from an active or passive $RC$ network, but the PLL offers some special features.

Very precise 90° phase shifters with accuracies better than ± 0.01° are easy to realize by PLLs. It is further possible to fabricate a phase shifter with a PLL that maintains a constant phase shift of 90° between reference and VCO output signals over a broad frequency range. Such a network is not easy to design when conventional filtering techniques are used.[30]

It is well known that a PLL using a type 1 or a type 2 PD (Table 2-1) exhibits a phase shift of 90° when operating at its center frequency. If an active loop filter having a pole at $s = 0$ is used (type 3 or 4 of Table 2-2), the phase shift is a constant 90° for all operating frequencies within the hold range. It should be realized, however, that very simple circuits are often used for a PD. Consequently, the phase shift between reference and output signals can deviate considerably from the theoretical value of 90°.

A very accurate 90° phase is obtained, however, if only a few elements are added to the basic PLL (Fig. 7-46). A precision multiplier cascaded by an integrator has been added. This arrangement is connected in parallel to the cascade consisting of the phase detector and the loop filter.

The average output signal of the precision multiplier is zero only if the signals $u_1$ and $u_2$ are *exactly* 90° out of phase, provided the signals are symmetrical. Furthermore the gain of the integrator is infinite at zero frequency. In practice, this means that the cascade consisting of the precision multiplier and integrator *overrides* the cascade consisting of the original PD and the loop filter, and the accuracy of the phase shift is given by the specifications of the precision multiplier alone.

If the VCO output signal is a square wave, an inexpensive transmission gate such as the CD 4016 can be used (CMOS) for the precision multiplier. Figure 7-47 shows an all-CMOS precision 90° phase shifter. All circuits are symmetrically powered by ± 7.5 V. The summation of the two error signals at the input of the VCO is done passively, as shown in Fig. 7-47.

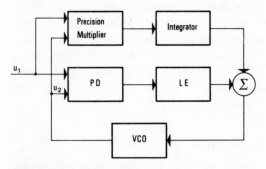

**Fig. 7-46  Block diagram of a precision 90° phase shifter.**

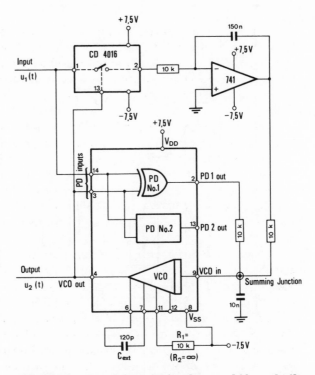

**Fig. 7-47 Precision 90° phase shifter built entirely from CMOS devices. Note that a CMOS operational amplifier can be substituted for the type 741 shown.**

The circuit of Fig. 7-47 also operates with input signals $u_1$ having a waveform other than a square wave. If $u_1$ has a dc component, however, it should be coupled capacitively to the transmission gate.

Precision phase shifters are the key elements in vector impedance (or admittance) measuring systems and in vector voltmeters. Figure 7-48 shows a vector impedance measuring circuit, which simultaneously measures the real $(R)$ and the imaginary $(X)$ parts of an unknown impedance

$$Z = R + jX$$

A sweep generator is used to generate the ac signal required for the impedance measurement. Hence the impedance can be measured in a broad frequency range. Because the ac frequency is proportional to a timing current, the momentary frequency is available as a dc output signal. As will be shown, signals $R$ and $X$ are also available as dc signals. Therefore the real and imaginary parts of an impedance can be plotted directly against frequency with an $XY$ recorder or an oscilloscope.

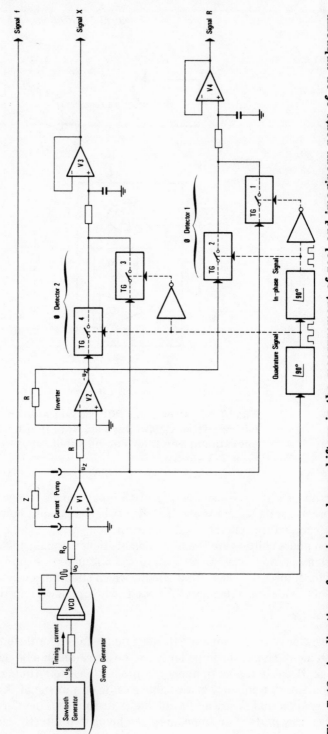

**Fig. 7-48 Application of precision phase shifters in the measurement of real and imaginary parts of an unknown impedance.**

Operational amplifier $V_1$ is used as a current source for the unknown impedance $Z$. The output $u_z$ of this amplifier is therefore proportional to $Z$. The real part of the impedance $Z$ is obtained by phase-sensitive detection of the signal $u_z$ with reference to a carrier signal, which is in phase with the current source or with the signal $u_0$ in Fig. 7-48 (which is equivalent to the current source). The imaginary part of $Z$ is obtained in the same way, but in this case the reference carrier is in quadrature with respect to the current source.

Both reference carriers are taken from the outputs of two precision 90° phase shifters, which are labeled $\underline{/90°}$ in the block diagram. Each of the two phase-sensitive detectors is constructed from two transmission gates, one of which is active (that is, ON) during the positive half-waves of the reference carrier; the other is ON during the negative half-waves.

A disadvantage of the circuit in Fig. 7-48 is that the unknown impedance cannot be grounded. This becomes possible, however, when an alternative current source, also known as a Howland circuit,[40] is used (Fig. 7-49). The remaining portion of the circuit is left unchanged.

The determination of the real part $R$ and the imaginary part $X$ is just one way of measuring an unknown impedance $Z$. As an alternative, the magnitude $|Z|$ and the phase $\phi$ could be measured. The magnitude and phase of an impedance $Z$ are defined by

$$Z = |Z|e^{j\phi}$$

A circuit measuring the magnitude and phase of an impedance $Z$ is shown in Fig. 7-50. The corresponding waveforms are plotted in Fig. 7-51. The first signal is the driving voltage $u_0$, which is in phase with the current driving the unknown impedance $Z$. Next the voltage $u_z$ across impedance $Z$ is shown. The phase $\phi$ is simply the phase shift between signals $u_0$ and $u_z$.

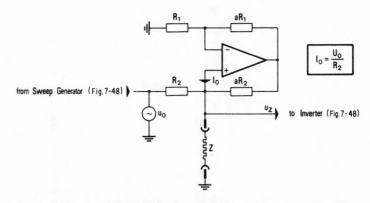

**Fig. 7-49  Modified driving circuit (compare with Fig. 7-48) enabling grounding of the impedance $Z$.**

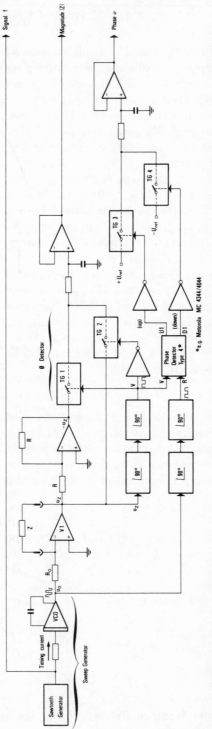

**Fig. 7-50  Application of precision phase shifters in the measurement of magnitude |Z| and phase φ of an unknown impedance.**

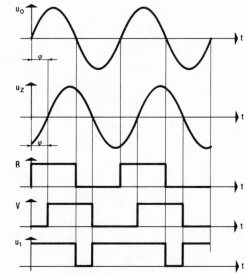

**Fig. 7-51 Waveforms relating to the circuit of Fig. 7-50.**

To measure the magnitude $|Z|$, the amplitude of signal $u_z$ has to be determined. For this purpose a square-wave reference signal $V$ is generated, which is exactly in phase with $u_z$. The $V$ signal is obtained by a cascade of two precision 90° phase shifters. Finally the magnitude $|Z|$ is obtained by the phase-sensitive detection of the signal $u_z$ with reference to the $V$ signal.

To determine the phase $\phi$, the phase shift between signals $u_0$ and $u_z$ has to be measured (see Fig. 7-51). This is most conveniently done by generating another square-wave reference signal $R$, which is in phase with signal $u_0$. As seen from Fig. 7-50, this signal is also obtained by a cascade connection of two precision 90° phase shifters.

The desired phase signal $\phi$ is simply the phase shift between signals $R$ and $V$. As shown in Fig. 7-50, $\phi$ is measured by a type 4 PD. (Table 2-1; refer also to Sec. 4-1-1). The UP and DOWN outputs of the phase detector are used to control two transmission gates, TG3 and TG4, which are assigned the weights $+1$ and $-1$, respectively. The phase signal $\phi$ is obtained by building the weighted average of the output signals of TG3 and TG4.

The same principle can also be used to measure an admittance $Y$. An admittance $Y$ can be defined either as

$$Y = G + jB$$

where $G$ is the real part or *conductance* and $B$ is the imaginary part or *susceptance,* or by magnitude and phase,

$$Y = |Y|e^{j\phi} = |Y| \underline{/\phi}$$

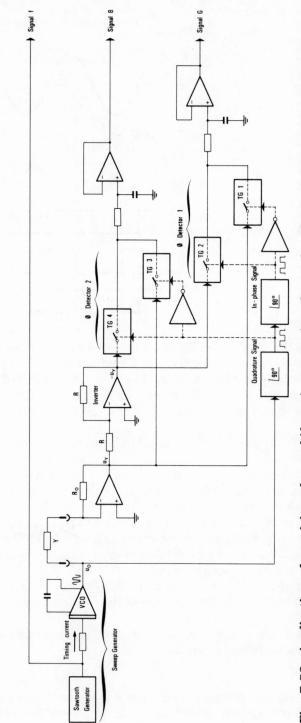

**Fig. 7-52** Application of precision phase shifters in measuring the real and imaginary parts of an unknown admittance.

236

A circuit measuring $G$ and $B$ is shown in Fig. 7-52. It operates in much the same manner as the circuit of Fig. 7-48. It also has the disadvantage that the unknown admittance cannot be grounded.

If grounding is desired, an alternative measuring circuit must be used (Fig. 7-53). It is built from four operational amplifiers. The first one (VI) acts as a voltage follower for the driving source $u_0$. Because the *current* flowing in $Y$ must be measured, it is sampled by the resistor $R_0$, which is in series with $Y$. Since no lead of $R_0$ is grounded, an *instrumentation amplifier*[41] is employed to amplify the differential signal across $R_0$. The instrumentation amplifier comprises three operational amplifiers: $V_2$, $V_3$, and $V_4$. Integrated instrumentation amplifiers are available in many versions and with various schemes.[12-14] An admittance-measuring curcuit evaluating magnitude and phase is easily obtained by combining the admittance-measuring circuit of Fig. 7-53 with the synchronous detectors shown in Fig. 7-50.

When analyzing the transfer function of an amplifier, a filter, or another device, one usually connects a swept sine-wave generator to its input and measures the magnitude and the phase of its output voltage. A device measuring the magnitude and the phase of a voltage is usually referred to as a *vector voltmeter*. The block diagram of a vector voltmeter is shown in Fig. 7-54. Its operation is analogous to that of the impedance-measuring circuit of Fig. 7-50.

The frequency range of the circuits discussed in this section is limited

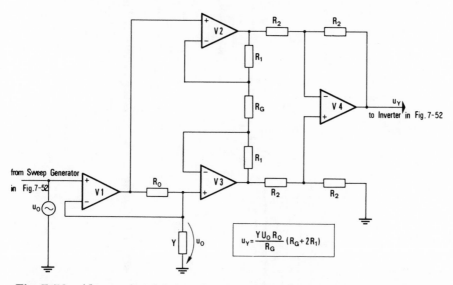

**Fig. 7-53** **Alternative driving circuit enabling grounding of the admittance $Y$.**

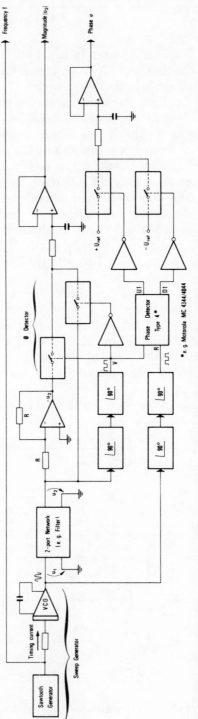

Fig. 7-54 Block diagram of a vector voltmeter built from precision phase shifters.

primarily by the slew rate of the operational amplifiers and also by the propagation delays of the transmission gates. The upper frequency limit is usually in the region of 100 kHz. Of course the bandwidth of the circuits could be extended by using video amplifiers instead of the slower operational amplifiers, diode quads instead of the transmission gates, and so on.

## 7-9 MOTOR-SPEED CONTROL WITH PLLs

Very precise motor-speed controls at low cost are possible using the PLL.[42-46] The advantages offered by the PLL technique become evident if the PLL motor-speed controls are first compared with conventional motor-speed controls. A classical scheme for motor-speed control is sketched in Fig. 7-55. The set point for the motor speed is given by the signal $u_1$. The shaft speed of the motor is measured with a tachometer; its output signal $u_2$ is proportional to the motor speed $n$. Any deviation of the actual speed from the set point is amplified by the servo amplifier whose output stage drives the motor. The gain of the servo amplifier is usually high but finite; to drive the motor, a nonzero error must exist.

Other sources of errors are nonlinearities of the tachometer and drift of the servo amplifier. A further drawback is the relatively high cost of the tachometer itself.

The PLL technique offers a much more elegant solution. Figure 7-56 is the block diagram of a PLL-based motor-speed control system. The entire control system is just a DPLL in which the VCO is replaced by a combination of a motor and optical tachometer, as shown in Fig. 7-56. The tachometer signal is generated by a fork-shaped optocoupler in which the light beam is chopped by a sector disk; the detailed circuit is shown in Fig. 7-57. The optocoupler is usually fabricated from a light-emitting diode (LED) and a silicon phototransistor. In order to obtain a clean square-wave output, the phototransistor stage is normally followed by a Schmitt trigger, such as a 74C14-type CMOS device.

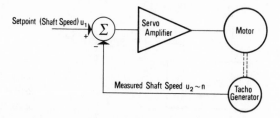

**Fig. 7-55 Block diagram of a conventional motor-speed control system.**

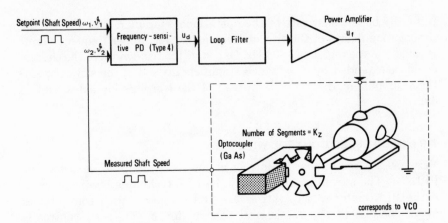

**Fig. 7-56 Block diagram of a motor-speed control system based on PLL techniques.**

The signal generated by the optocoupler has a frequency proportional to the speed of the motor. Because the phase detector compares not only the frequencies $\omega_1$ and $\omega_2$ of the reference and the tachometer signals but also *their phases,* the system settles at *zero-velocity error.* To enable locking at every initial condition, a phase- and frequency-sensitive PD must be used, such as the type 4 PD (Table 2-1).

To analyze the stability of the system, the transfer functions of all blocks in Fig. 7-56 must be known. The transfer functions of the PD and of the loop filter are already given [refer to Eq. (3-1)], but the transfer function of the motor-tachometer combination still must be determined.

If the motor is excited by a voltage step of aplitude $u_f$, its angular speed $\omega(t)$ will be given by

$$\omega(t) = K_m u_f[1 - \exp(-t/T_m)] \tag{7-32}$$

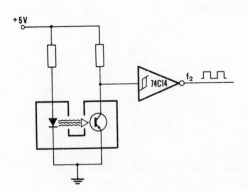

**Fig. 7-57 Circuit of the optical tachometer generator shown in Fig. 7-56.**

240

where $K_m$ is the proportional gain and $T_m$ is the mechanical time constant of the motor. The step response of the motor is plotted on the right in Fig. 7-58. Equation (7-32) indicates that $\omega$ will settle at a value proportional to $u_f$ after some time. Applying the Laplace transform to Eq. (7-32) yields

$$\Omega(s) = U_f(s) \frac{K_m}{1 + sT_m} \tag{7-33}$$

The phase angle $\phi$ of the motor is the time integral of the angular speed $\omega$. Therefore we get for its Laplace transform $\Phi(s)$ the expression

$$\Phi(s) = U_f(s) \frac{K_m}{s(1 + sT_m)} \tag{7-34}$$

The sector disk shown in Fig. 7-56 has $K_z$ teeth. This implies that the phase of the tachometer signal is equal to phase $\phi$ multiplied by $K_z$. Consequently we obtain for $\Theta_2(s)$

$$\Theta_2(s) = \frac{K_m K_z}{s(1 + sT_m)} U_f(s) \tag{7-35}$$

The transfer function $H_m(s)$ of the motor is therefore given by

$$H_m(s) = \frac{K_m K_z}{s(1 + sT_m)} \tag{7-36}$$

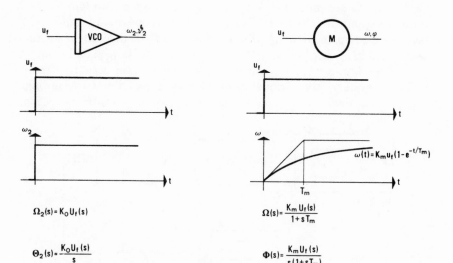

**Fig. 7-58** The step response of an ordinary VCO compared with that of a motor. Note that the VCO is a first and the motor a second-order system.

The motor is evidently a second-order system, whereas the VCO [according to Eq. (3-1)] was a first-order system only. In Fig. 7-58 the transient response of the motor is compared with that of an ordinary VCO. The motor-speed control system of Fig. 7-56 is therefore a third-order system. The mathematical model of the control system can now be plotted (Fig. 7-59). The servo amplifier is supposed to be a zero-order gain block with proportional gain $K_a$. The poles of this amplifier normally can be neglected because they are at much higher frequencies than the poles of the motor. The closed system has three poles. Therefore, a filter with a zero (type 2 or type 4 according to Table 2-2) must to be specified for the loop filter, otherwise the phase of the closed-loop transfer function would exceed 180° at higher frequencies and the system would become unstable.

The individual blocks in Fig. 7-59 can be combined into fewer blocks, a result which yields the simpler block diagram of Fig. 7-60. In this system the transfer function of the forward path is defined by $G(s)$, whereas the transfer function of the feedback network (motor) is given by $H(s)$.

When a motor-speed control system is designed practically, some parameters are initially given, such as the motor parameters $K_m$ and $T_m$ or the number of teeth $K_Z$ of the sector disk. The remaining parameters ($K_a$ and $\tau_2$) then have to be chosen for best dynamic performance and maximum stability of the system. There are many ways to solve this problem. It is possible to calculate the hitherto unspecified parameters by purely mathematical methods. In design engineering, however, more practical methods are preferred. We will therefore discuss two familiar techniques: design with the Bode diagram and the root-locus-plot method.[3,4]

The Bode-plot method is discussed first. A Bode diagram of the function $G(j\omega)H(j\omega)$ is plotted, and the phase margin is estimated from this plot. If the phase margin is considered too low or even becomes negative (unstable system), the parameters of the system are adjusted accordingly. Figure 7-61 demonstrates how the Bode plot of $G(j\omega)H(j\omega)$ is con-

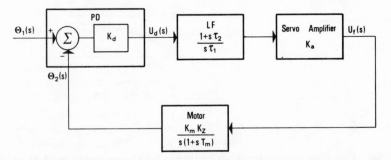

**Fig. 7-59  Mathematical model of the motor-speed control system of Fig. 7-56.**

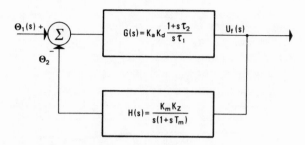

**Fig. 7-60  Condensed mathematical model of the motor-speed control system.**

structed. $G(j\omega)H(j\omega)$ is given by

$$G(j\omega)H(j\omega) = K_d K_a K_m K_Z \frac{1 + j\omega\tau_2}{(j\omega)^2(1 + j\omega T_m)} \qquad (7\text{-}37)$$

This can be decomposed into three simple terms,

$$G(j\omega)H(j\omega) = K(1 + j\omega\tau_2) \frac{1}{1 + j\omega T_m} \frac{1}{j\omega^2} \qquad (7\text{-}38)$$

In this expression all individual gain factors are combined into $K$,

$$K = \frac{K_d K_a K_m K_Z}{\tau_1}$$

The easiest way to construct a Bode diagram is to first plot the Bode diagrams of the three terms in Eq. (7-38). In order to assign numerical values to the Bode plot, we must choose some initial values for $K$, $\tau_2$, and $T_m$. We arbitrarily define

$$K = 300 \ (\triangleq 50 \text{ dB})$$

$$\tau_2 = 0.33 \text{ s}$$

$$T_m = 50 \text{ ms}$$

The Bode diagram for the term $(1 + j\omega\tau_2)$ is shown in Fig. 7-61a. The curve is horizontal below the corner frequency $1/\tau_2$ and then rises with a slope of $+20$ dB per decade. The phase approaches zero at very low frequencies and $+\pi/2$ at very high frequencies; it is exactly $+\pi/4$ at the corner frequency. The slope of the phase in the proximity of the corner frequency is approximately $+\pi/4$ per decade of frequency. With these simple rules of thumb, we can manually plot the Bode diagram with sufficient accuracy. If a higher precision is desired, the phase can be read from the graph in Fig. 7-62, which plots phase against normalized frequency $\omega/\omega_0$ for first-order lag and lead systems.

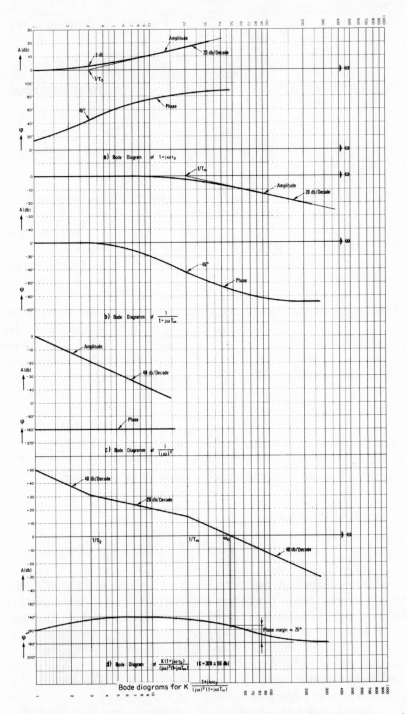

**a)** Bode Diagram of $1 + j\omega\tau_2$

**b)** Bode Diagram of $\dfrac{1}{1 + j\omega T_m}$

**c)** Bode Diagram of $\dfrac{1}{(j\omega)^2}$

**d)** Bode Diagram of $\dfrac{K(1 + j\omega\tau_2)}{(j\omega)^2(1 + j\omega T_m)}$    ($K = 300 \triangleq 50$ db)

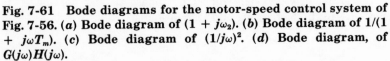

Bode diagrams for $K\dfrac{1 + j\omega\tau_2}{(j\omega)^2(1 + j\omega T_m)}$

**Fig. 7-61 Bode diagrams for the motor-speed control system of Fig. 7-56. (a) Bode diagram of $(1 + j\omega_2)$. (b) Bode diagram of $1/(1 + j\omega T_m)$. (c) Bode diagram of $(1/j\omega)^2$. (d) Bode diagram, of $G(j\omega)H(j\omega)$.**

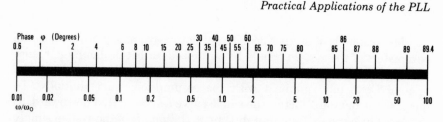

**Fig. 7-62** **Gain/phase plot for a first-order lag network. The phase $\phi$ is plotted against the normalized frequency $\omega/\omega_0$.**

The Bode diagram for $1/(1 + j\omega T_m)$, the second term of Eq. (7-36), is plotted in an analogous way and is shown in Fig. 7-61$b$. Here the amplitude rolls off at $-20$ dB per decade above the corner frequency $1/T_m$, and the phase is between 0 and $-\pi/2$. The Bode diagram for $1/(j\omega^2)$, the third term of Eq. (7-36), is shown in Fig. 7-61$c$. The amplitude rolls off at $-40$ dB per decade, and the phase is a constant $-\pi$.

Superposition of all three curves yields the Bode plot of the function $G(j\omega)H(j\omega)$, which is shown by Fig. 7-61$d$. The Bode plot drastically demonstrates the effect of the zero at the frequency $1/\tau_2$. If this zero were nonexistent, the gain would roll off at $-60$ dB per decade and the overall phase would become $-3\pi/2$ ($-270°$), which means that the system would be unstable. With the assumptions made ($K = 300$, $\tau_2 = 0.33$ s, and $T_m = 50$ ms), the phase margin becomes approximately $25°$, which indicates that the system is stable. This statement alone is not sufficient, however, because we do not yet know whether the system is overdamped or underdamped, and how large the bandwidth of the closed system actually is. The closed-loop bandwidth $\omega_G$ is obtained easily by the intercept of the amplitude curve with the $\omega$ axis, $\omega_G \approx 44$ s$^{-1}$.

The damping factor $\zeta$ can be obtained by an approximation indicated by Osborne in Ref. 3,

$$\zeta \approx \frac{\alpha}{3n}$$

where $\alpha$ is the phase margin in degrees (!) and $n$ is the slope of the gain curve in dB per decade at the closed-loop bandwidth $\omega_G$. By definition $n$ is a positive number if the slope is negative. From the Bode diagram of Fig. 7-61$d$ we read approximately

$\alpha \approx 25°$

$n \approx 40$ dB per decade

Hence

$\zeta \approx 0.2$

The system is underdamped. To get better dynamic performance, the parameters of the system should be optimized for better damping. Figure 7-63 demonstrates how this can be done. Curve 1 in this diagram is the original Bode plot obtained with the assumptions made initially. To increase the damping factor, the whole curve can be shifted downward, as indicated by curve 2. To obtain this curve, the gain $K$ of the system simply must be reduced (in this example by 14 dB). This decreases the closed-loop bandwidth from 44 s$^{-1}$ to approximately 20 s$^{-1}$. The slope of the amplitude curve is now about $-30$ dB per decade at the "new" closed-loop bandwidth, and the phase is estimated to be about $-135°$. Then the phase margin is approximately $\alpha \approx 45°$. With these adjusted values the damping factor becomes

$$\zeta \approx \frac{45}{3.30} = 0.5$$

This is an acceptable figure. The same effect could have been obtained by leaving the gain $K$ unchanged but by displacing the zero of the loop filter from the original frequency $1/\tau_2 = 3$ s$^{-1}$ to the new frequency $1/\tau_2 = 15$ s$^{-1}$. When this is done, curve 3 in Fig. 7-63 is obtained. Curves 2 and 3 are almost identical above the corner frequency $1/T_m$, and the dynamic performance of both variants will be about the same.

In the case of curve 3 the gain at lower frequencies is higher than it is for curve 2, which means that at lower frequencies the dynamic errors are smaller for the compensation according to curve 3.

We cannot generalize, however, that shifting the loop filter zero to higher frequencies necessarily improves the dynamic performance of the control system. This can even have an adverse effect on the system, as shown by the example in Fig. 7-64. Here the onset of the loop filter zero is at such a high frequency (90 s$^{-1}$) that the overall gain has already rolled off at $-60$ dB per decade, which makes the system unstable.

Let us now discuss the root-locus method. Here the stability of a system is analyzed by considering the location of the poles of the transfer function according to Eq. (7-37). The location of the poles is obtained by setting

$$G(s)H(s) = -1 \qquad (7\text{-}39)$$

and solving for the roots of this equation. Because our system is a third-order system, three roots are obtained. The location of the roots depends on the parameters of the system such as $K$, $T_m$, and $\tau_2$. In the root-locus method all parameters except one are chosen constant, and the remaining parameter ($K$) is variable, usually between zero and infinity.

For each value of the varied parameter, a set of three roots is obtained. In a third-order system the roots of the system migrate on three

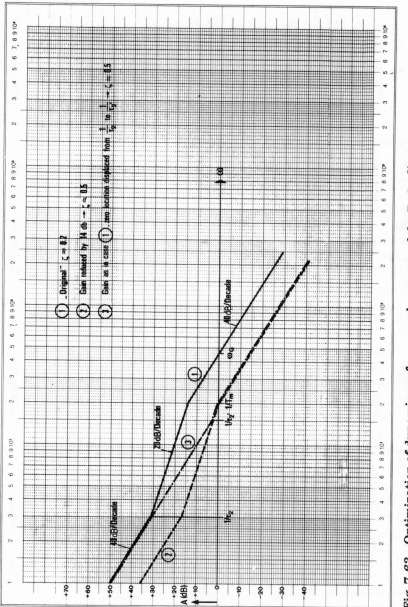

Fig. 7-63  Optimization of dynamic performance by means of the Bode diagram.

247

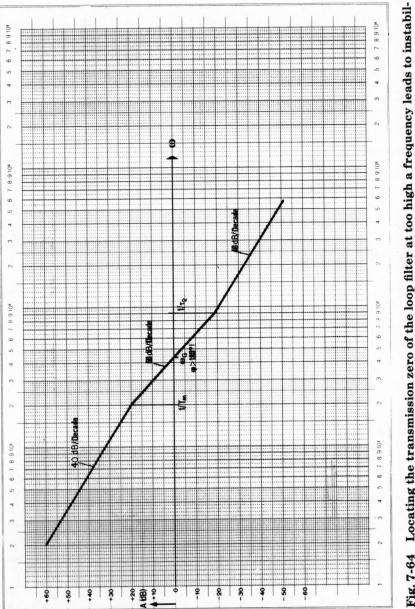

**Fig. 7-64 Locating the transmission zero of the loop filter at too high a frequency leads to instability in this case.**

different trees, as shown in Fig. 7-65. The system is said to be unconditionally stable if the roots stay in the left half of the $s$ plane for any value of the varied parameter. The system is called conditionally stable if the roots are in the left half for a limited range of the varied parameter only.

Returning to our motor-speed control system, a root locus was plotted for $T_m = 50$ ms and $\tau_2 = 0.33$ s (Fig. 7-65$a$), with $K$ being the variable parameter. It was found that the system is unconditionally stable, since the roots are always in the left half of the $s$ plane. One root is real, the other two are a complex conjugate pair. Theoretically the roots are obtained by solving the third-order equation (7-39). This would be extremely cumbersome. Some practical hints for shortening the work are given in Ref. 4; however they will not be discussed here. Nevertheless, the root-locus method is much clumsier than the Bode-diagram technique.

If the stability of the system needs to be investigated for another set of the parameters $T_m$ and $\tau_2$, a new root locus must be constructed. This is illustrated in Fig. 7-65$b$, which shows the root locus for the case $T_m = 50$ ms and $\tau_2 = 0.08$ s. Moreover, it is very difficult to predict the values of $T_m$ and $\tau_2$ for which the system will become unstable. For all these reasons, the Bode-diagram technique is the preferred method.

In addition to the hardware circuits discussed in this section, software has also been used in the design of PLL motor-speed control systems.[46]

## 7-10 THE SYNCHRONOUS FILTER

The synchronous filter is related to the tracking filter described in Sec. 7-1. The tracking filter has been defined as a network producing an output signal $u_2(t)$ which is phase-locked to the input signal $u_1(t)$. With the tracking filter the "reconstructed" signal $u_2$ could be used for a number of different purposes, such as for synchronous demodulation of the input signal (refer to Figs. 7-22–7-25), or for measuring the frequency of the (noisy) input signal. The synchronous filter goes a step further: it permits reconstruction of the waveshape of the input signal $u_1$ itself.[47] The bandwidth of the synchronous filter can be made arbitrarily narrow, as will be seen in this section. Consequently, signals buried in noise can be reshaped with high accuracy.

The block diagram of a synchronous filter is shown in Fig. 7-66$a$; its principle of operation is best understood by looking at the waveforms in Fig. 7-66$b$. First an ordinary tracking filter is used to generate an output signal $u_2$, which is phase locked to the (noisy) input signal $u_1$. The PD used in the system is supposed to be linear (type 1 in Table 2-1). Because there is a frequency divider between the VCO and the PD, the VCO will oscillate at $k$ times the frequency of the input signal. In the example shown $k = 8$. In the circuit of Fig. 7-66$a$, $u_2$ is used to shift a logic 1

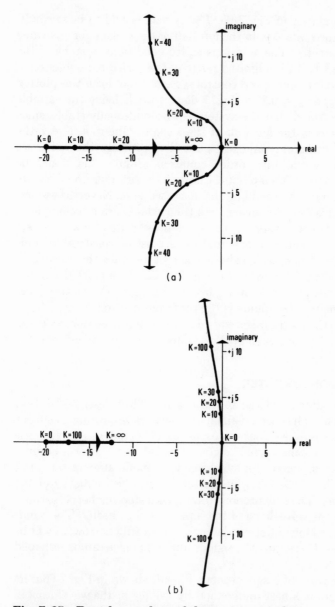

**Fig. 7-65  Root-locus plots of the motor-speed control system for two values of $\tau_2$. ($T_m = 50$ ms) ($a$) $\tau_2$ = 0.33 s. ($b$) $\tau_2$ = 0.08 s.**

continuously through an 8-bit shift register; some additional logic, not shown in the drawing, ensures that no more than one stage of the shift register can contain a 1.

The outputs $Q_1$ to $Q_8$ of the shift register control the two groups of eight analog switches $S_{1a}$ to $S_{8a}$ and $S_{1b}$ to $S_{8b}$. Analog switch $S_{1a}$ is closed periodically during the first eighth of an input signal period, $S_{2a}$ is ON during the second eighth, and so on. Consequently the signal across capacitor $C_1$ is equal to the input signal averaged over the first eighth of the signal period. The set of eight capacitors represents the filtered input signal resolved into eight discrete samples. As seen from Fig. 7-66a, the eight analog switches $S_{1b}$ to $S_{8b}$ connect successively the eight samples to the input of a voltage follower. The output signal of this amplifier is therefore a staircase approximation of the input signal $u_1$. The resolution of the synchronous filter can be improved by increasing the number $k$ of the shift register stages and the analog switches.

The bandwidth of the synchronous filter can be calculated from its transfer function. The transfer function[12] is given by

$$H(s) = \frac{1 - \exp(sT/k)}{s\tau[1 - \exp(-T/k\tau)e^{-sT}]} \qquad (7\text{-}40)$$

where $T$ = period of the input signal
  $k$ = number of samples taken
  $\tau$ = time constant of the $RC$ filters in Fig. 7-66a, $\tau = RC_1$
    $= RC_2 = \cdots$

The magnitude $|H(j\omega)|$ of this transfer function is plotted against $\omega$ in Fig. 7-66c. The transfer function is found to be periodic with the frequency.

Input signals having frequencies $f_s$, $2f_s$, $3f_s$, ... ($f_s = 1/T$) will also be passed through the filter, but signals having frequencies $kf_s$, $2kf_s$, ... will be suppressed. For these frequencies the averaging period in the synchronous filter is an integer multiple of the signal period; consequently the average value is 0.

The bandwidth of the synchronous filter (one-sided 3-dB bandwidth, see Fig. 7-66c) is given by[17]

$$\Delta\omega = \frac{1}{k\tau} \qquad (7\text{-}41)$$

and hence can be made very small by just one single $RC$ combination.

The cutoff characteristic of the synchronous filter can be made even steeper either by cascading a number of first-order switched filters (Fig. 7-66d) or by using a number $k$ of higher-order low-pass filters as shown

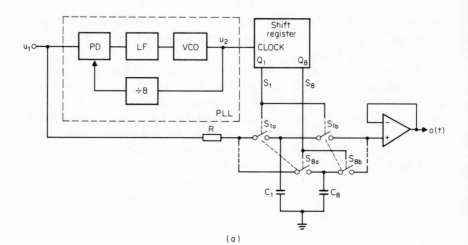

(a)

(b)

**Fig. 7-66  A synchronous filter. (a) Block diagram. (b) Corresponding waveforms. (c) Transfer function $H(j\omega)$. (d) Filter section of the synchronous filter built from a cascade of passive first-order, low-pass filters. (e) Filter section of the synchronous filter built from higher-order low-pass filters (Butterworth characteristic).**

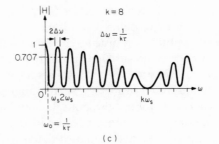

(c)

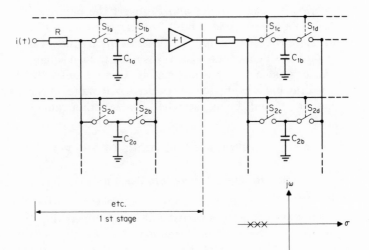

etc.

1 st stage

(d)

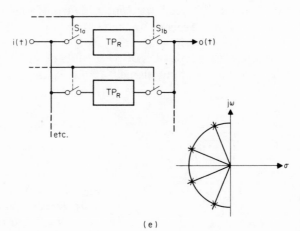

(e)

in Fig. 7-66$e$. In the latter case, active filters can be applied, such as active *RC* Chebyshev or elliptic low-pass filters.

The major application of the synchronous filter is the reconstruction of signals buried in noise. One interesting application is synchronous demodulation in AM receivers.[29] It has been shown (by the example of the AM receiver illustrated in Fig. 7-23) that analog multipliers are prone to distortion when used for synchronous demodulation. The synchronous filter shows superior performance because its demodulating part consists of linear elements only. To demodulate a signal synchronously, it is sufficient to work with two samples and to invert the polarity of one of them. A synchronous demodulator built from a synchronous filter is shown in Fig. 7-67. Here the VCO simply oscillates at the frequency of the input signal. The output signal of the VCO is assumed to be a square wave. It turns on a pair of analog switches alternately. Consequently the capacitor shown in the drawing is averaging the positive half-waves of the input signal with a gain of $+1$ and the negative half-waves with a gain of $-1$.

## 7-11  MISCELLANEOUS APPLICATIONS OF THE PLL

### 7-11-1 Analog-to-Digital Converters Using a VCO

The VCO is basically a voltage-to-frequency converter. When a few logic elements are added, a low-cost ADC is obtained (Fig. 7-68). The pulses generated by the VCO are summed in an UP counter over a fixed time interval $T_0$. At the end of the time interval $T_0$ the content of the counter is thus proportional to the input signal *averaged* over $T_0$. As seen from the waveforms in Fig. 7-68, the content is transferred periodically at the end of $T_0$ to a buffer register by the LOAD command. The counter is reset immediately thereafter.

This integrating type of ADC is capable of suppressing noise signals superimposed on the input signal.[25] If the signal contains hum coupled

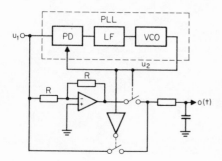

**Fig. 7-67 The synchronous detector is a special case of the synchronous filter.**

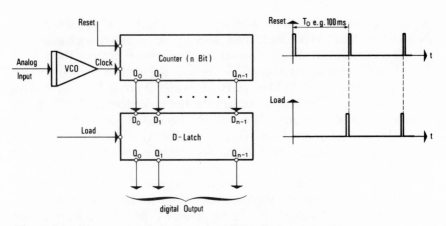

**Fig. 7-68   Analog-to-digital converter built from a VCO.**

from the ac line (60 Hz), this disturbance is canceled fully when $T_0$ is set at an integer multiple of a line period (16⅔ ms).

The voltage-to-frequency converter is a relatively slow type of ADC because it updates the digital output signal at a rate given by the interval $T_0$, which is chosen quite long in most cases.

### 7-11-2 Stereo Decoders

There are different types of stereo decoders in use. The *matrix stereo decoder*[23,48] is discussed in this section.

In stereo radio two signal channels are transmitted: the $L + R$ (left + right) and the $L - R$ (left − right) channels. On the transmitter site, the two audio signals $L$ and $R$ are first converted into the composite signal $u_c$. The frequency spectrum of $u_c$ is shown in Fig. 7-69. First an $L + R$

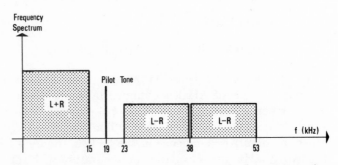

**Fig. 7-69 Frequency spectrum of the composite stereo signal $u_c$.**

255

signal is generated by simply adding $L + R$. The $L + R$ signal is a baseband signal (0–15 kHz), for its frequency spectrum is left unchanged.

Next an $L - R$ signal is generated. To separate it from the $L + R$ signal, the $L - R$ signal is amplitude-modulated with an auxiliary carrier having a frequency of 38 kHz. DSB-AM with suppressed carrier is used. At an audio signal bandwidth of 15 kHz, the up-converted $L - R$ signal covers a frequency range of 23 to 53 kHz. Instead of the auxiliary carrier (38 kHz), a pilot tone of half that frequency (19 kHz) is added to the spectrum. In mono transmission the pilot tone and the $L - R$ signal are switched off.

At the transmitter site the composite signal $u_c$ is finally frequency-modulated onto an RF carrier. These two different modulations, (1) DSB-AM of the $L - R$ signal onto the (suppressed) 38-kHz auxiliary carrier and (2) frequency modulation of the composite signal onto an RF carrier within the frequency range of 88–108 MHz, must not be confused.

In stereo transmission the composite signal $u_c(t)$ is given by

$$u_c = (L + R) + (L - R) \cos \omega_s t + u_p \cos \frac{\omega_s}{2} t$$

where $\omega_s$ is the angular frequency of the (suppressed) auxiliary carrier, $\omega_s = 2\pi \cdot 38\,000 \text{ s}^{-1}$, and $u_p$ is the amplitude of the pilot tone.

The stereo receiver has to check first whether or not the pilot tone is present. In the absence of the pilot tone, the $L + R$ signal has to be distributed evenly between the $L$ and $R$ channels. If the pilot tone is on, however, the receiver will have to decode the individual channels from the composite signal. The block diagram of a matrix stereo decoder is shown in Fig. 7-70, and the corresponding waveforms are plotted in Fig. 7-71.

The matrix decoder can be subdivided into three partial blocks:

1. A linear PLL generating two 19-kHz square-wave signals which are in quadrature

2. An in-lock detector

3. The intrinsic decoder circuit

The VCO of the PLL is oscillating at a center frequency of 76 kHz. The auxiliary carrier (38 kHz) is obtained by a $\div 2$ divider within the PLL. Two further $\div 2$ dividers ($JK$ flip-flops) are used to generate two 19-kHz square waves. One of these $JK$ flip-flops counts the positive-going, the other the negative-going edge of the 38-kHz square wave. Hence, one of the two 19-kHz square waves is in quadrature with the pilot tone and is used as an input to the PD, while the other is in phase with the pilot tone and is used as an input to the in-lock detector. When the pilot tone is

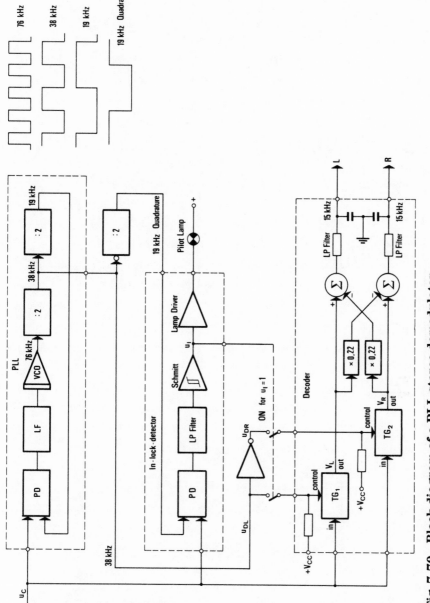

**Fig. 7-70  Block diagram of a PLL stereo demodulator.**

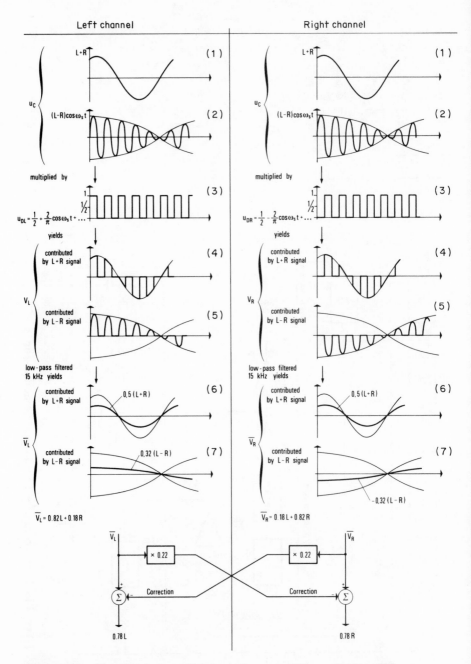

**Fig. 7-71 Waveforms explaining the operation of the PLL stereo demodulator in Fig. 7-70.**

detected, the output signal $u_I$ of the Schmitt trigger is a logic 1, and the pilot lamp is lit. If $u_I$ is 1, the two analog switches $S$ are by definition in the ON state.

Let us first consider the MONO mode ($u_I = 0$). Both analog switches $S$ are OFF. Therefore both transmission gates $TG_1$ and $TG_2$ are in the ON state, which means that the $L + R$ signals appear at both outputs $L$ and $R$ of the decoder with an identical amplitude. Note that no $L - R$ signal exists in this mode.

Next we consider the STEREO mode. Because now $u_I = 1$, the transmission gates $TG_1$ and $TG_2$ are switched periodically by the 38-kHz auxuliary carrier. More precisely they are switched in a push-pull manner because the control signals $u_{DL}$ and $u_{DR}$ are of opposite phase.

The operation of the matrix decoder is best explained by the waveforms of Fig. 7-71. Consider the left channel. The composite signal $u_c$ is shown decomposed into the $L + R$ signal (trace 1) and the $L - R$ signal (trace 2). When passing through the transmission gate $TG_1$, both are multiplied by the signal $u_{DL}$ (trace 3). As a result, the baseband signal $L + R$ is chopped with a frequency of 38 kHz (trace 4). The $L - R$ signal, however, is demodulated synchronously (trace 5). The sum of traces 3 and 5 forms the signal $V_L$. The low-pass filter shown in Fig. 7-70 builds the average value of the signal $V_L$, denoted by $\overline{V_L}$. The chopped $L + R$ signal contributes an average of $0.5(L + R)$ to the signal $\overline{V_L}$ (see traces 4 and 6), and the synchronously demodulated $L - R$ signal contributes an average of $0.32(L - R)$ to the signal $\overline{V_L}$ (see traces 5 and 7). (Note that the average value of a synchronously rectified, half-wave sine signal is $\frac{1}{2} \, 2/\pi \approx 0.32$ times the peak amplitude.) Consequently the averaged signal $\overline{V_L}$ is given by

$$\overline{V_L} = 0.5(L + R) + 0.32(L - R) = 0.82L + 0.18R$$

In a similar way it can be shown that $\overline{V_R}$ of the right channel is given by

$$\overline{V_R} = 0.5(L + R) - 0.32(L - R) = 0.18L + 0.82R$$

To obtain two signals proportional to $L$ or $R$ alone, a correcting network must be added, as shown in the lower part of Fig. 7-70. By simple computations we obtain

$$0.78 \begin{pmatrix} L \\ R \end{pmatrix} = \begin{pmatrix} 1 & -0.22 \\ -0.22 & 1 \end{pmatrix} \begin{pmatrix} \overline{V_L} \\ \overline{V_R} \end{pmatrix} \tag{7-42}$$

so that a signal depending solely on on the $L$ channel is easily obtained by subtracting 0.22 times the right channel from the left one.

A number of ICs containing entire stereo decoder circuits are available (Table 6-1).

### 7-11-3 Color Subcarrier Synchronization in Television Receivers

In black-and-white television only the luminance signal $Y$ is transmitted. This signal represents the intensity of each dot of the television frame. With color television the color information (the so-called chrominance signal) has to be transmitted as well. Each color can be decomposed into the three primary colors red, blue, and green ($R$, $B$, and $G$), respectively. The sum of $R$, $B$, and $G$ is equal to the luminance signal $Y$. Because the luminance signal is transmitted anyway, only two chrominance signals need be added to the signal: $R$ and $B$. Actually the $R$ and $B$ signals themselves are not transmitted, but the equivalent signals $R - Y$ and $B - Y$.

There are a number of different systems of adding color information to the luminance signal. The oldest one, the American NTSC (National Television System Committee) system (often called ironically "Never The Same Color") uses a color subcarrier with a frequency of approximately 3.54 MHz to carry the color information. The two chrominance signals have to be modulated onto this subcarrier. QAM with suppressed carrier is used (refer to Sec. 7-3), for this method allows the transmission of two independent signals by the same carrier. Consequently the transmitter first generates an in-phase and a quadrature-phase color subcarrier. The $B - Y$ chrominance signal is modulated onto the (suppressed) in-phase subcarrier, and the $R - Y$ signal onto the (suppressed) quadrature subcarrier. The phase of the color spectrum is shown in Fig. 7-72. This diagram demonstrates that any particular color is obtained by a geometrical addition of the two components $R - Y$ and $B - Y$.

The resulting chrominance signal is therefore a vector. The phase of this vector defines the hue, the length of the vector the chroma of a particular color.

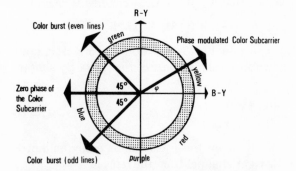

**Fig. 7-72  Color is represented as a vector in color television. The phase angle corresponds to hue, the length of the vector to chroma.**

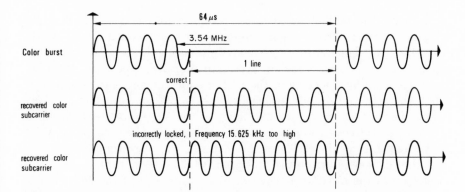

**Fig. 7-73  Reconstruction of the color subcarrier in a television receiver.**

If the angular frequency of the color subcarrier is denoted by $\omega_F$, the two components of the chrominance signal can be written as

$$(B - Y)(t) = (B - Y) \exp (j\omega_F t)$$

$$(R - Y)(t) = (R - Y) \exp \left[ j \left( \omega_F t + \frac{\pi}{2} \right) \right]$$

$$= j(R - Y) \exp (j\omega_F t)$$

The receiver now has to decode the individual chrominance signals $R - Y$ and $B - Y$. This is done by synchronous demodulation of the composite chrominance signal by means of two reconstructed color subcarriers, which are in quadrature. To reconstruct the color subcarrier, the receiver should "know" the phase of the color subcarrier as generated by the transmitter. (Remember that the color subcarrier itself is suppressed.) For this reason the transmitter periodically inserts the so-called color burst in the pauses between subsequent rows of the video signal (refer to Fig. 7-73, top trace). The color burst is thus repeated at a rate of 15.625 kHz.

At the receiver site a PLL system is used to reproduce a *continuous* color subcarrier, which is phase-locked to the color burst. Synchronizing a PLL onto the color burst does not appear as a difficult task, but a small problem should not be overlooked. In contrast to most PLL applications considered earlier, this PLL has to lock onto a signal which is available only for short periods of time. The second trace in Fig. 7-73 represents the case where the PLL has locked onto the correct frequency of 3.54 MHz. But assume now that the PLL has locked onto a frequency which is offset by just 15.625 kHz from the correct value (see bottom trace in Fig. 7-73). (Note that the drawing is not to scale; the relative duration of the burst is exaggerated here.) When this is the case, the VCO of the PLL is producing just one oscillation more than required during one line of the video signal. When the "next" color burst appears (right side of Fig. 7-73), the VCO output signal appears perfectly in phase with the burst; hence the phase detector does not see any reason to alter the frequency of the VCO.

To avoid false locking of the PLL, its lock range must be limited to half the row (horizontal) frequency, that is, to approximately $\pm 7$ kHz. This means that the center frequency of the VCO must be accurate to about 1 kHz. In other words, a quartz oscillator must be used for the VCO. The frequency of the oscillator can be tuned by a varactor diode.

Let us now return to the American NTSC system. In this system the color subcarrier has always the same phase, namely, 180°, as shown in Fig. 7-72. If envelope distortion occurs between transmitter and receiver, the chrominance signal can show a phase error with respect to the color subcarrier, which results in an unwanted alteration of the hue. This disadvantage is circumvented in the German PAL (phase alternation line) system by changing the polarity of the $R - Y$ signal in each alternating line. The chrominance signal $R - Y$ is thus transmitted with its original phase in the odd lines and with opposite polarity in the even lines. For the $R - Y$ signal we therefore get

$$(R - Y)(t)_{1,3,5, \ldots} = +j(R - Y) \exp (j\omega_F t)$$

$$(R - Y)(t)_{2,4,6, \ldots} = -j(R - Y) \exp (j\omega_F t)$$

The composite chrominance signal for PAL can then be written for odd lines as

$$U_F(t)_{1,3,\ldots} = (B - Y)(t) + j(R - Y)(t)$$
$$= [(B - Y) + j (R - Y)] \exp (j\omega_F t)$$

and for even lines as

$$U_F(t)_{2,4, \ldots} = (B - Y)(t) - j(R - Y)(t)$$
$$= [(B - Y) - j(R - Y)] \exp (j\omega_F t)$$

In the receiver the $B - Y$ signal is obtained by *adding* the $B - Y$ signal of the current line to the $B - Y$ signal of the corresponding dot in the previous line.

$$U_F(t)_i + U_F(t)_{i-1} = 2(B - Y) \exp (j\omega_F t)$$

It will be shown later how the delayed signal is obtained. The $R - Y$ signal is obtained by *subtracting* the $R - Y$ signal of the current line from the $R - Y$ signal of the corresponding dot in the previous line; for the resulting $R - Y$ signal we obtain

$$U_F(t)_i - U_F(t)_{i-1} = \begin{cases} +2j(R - Y) \exp (j\omega_F t) & \text{for } i = \text{odd} \\ -2j(R - Y) \exp (j\omega_F t) & \text{for } i = \text{even} \end{cases}$$

where $i$ denotes the line number. We see from the above equation that the $R - Y$ signal has inverted polarity in the even lines. In the PAL system this signal must be inverted in each even line. This is done by the so-called PAL switch, but the receiver first must "know" whether the current line is an even or an odd line. This is done in PAL by additional PM of the color subcarrier, as shown in Fig. 7-72. As mentioned earlier, the phase of the color subcarrier is a constant 180° for NTSC. In PAL, however, the subcarrier is additionally phase-shifted by $-45°$ in the even lines and by $+45°$ in the odd lines. The operation of the PAL decoder is explained in Fig. 7-74. First the color burst is used to synchronize the PLL. The noise bandwidth of the PLL is made so narrow that the PLL does *not* track the phase of the subcarrier, but is settles on its *average phase,* that is, at 180° (as was the case with NTSC). Consequently the PD "sees" a negative phase error in the even lines and a positive phase error in the odd lines. As shown in Fig. 7-74, the output signal of the PD is used to alternately set and reset a flip-flop. The $Q$ output of this flip-flop is used to control the PAL switch, thus correcting the phase of the $R - Y$ signal in the even lines.

The delayed signal (previous line) is obtained by a delay line having a delay corresponding to the duration of one line, approximately 64 $\mu$s. As shown in Fig. 7-74, the individual chrominance signals are finally obtained by synchronous detection.

### 7-12 IMPULSE-SYNCHRONOUS CIRCUITS USING THE PLL

This last section on PLL applications covers a special type of PLL, the impulse-synchronous PLL. The applications considered so far have been non-impulse-synchronous ones. The term impulse-synchronous is best explained by looking again at the motor-speed control system shown in Fig. 7-56.

When the set point of the shaft speed is kept constant and when the motor has reached its final speed, the system is perfectly synchronized, that is, $\omega_1 = \omega_2$. When the set point is abruptly changed, however, the synchronization is lost temporarily, which means that the overall number of pulses generated by the reference signal is not equal to the overall number of pulses generated by the tachometer. If the setpoint is increased, the motor may "miss" a number of pulses before acquiring synchronism again; if on the other hand the set point is suddenly decreased, the motor may overshoot and produce some extraneous pulses. This PLL system is said to be not impulse-synchronous.

Mathematically speaking, we could say that the type 4 PD used in this system is operating modulo $2\pi$; i.e., the phase detector cannot differ-

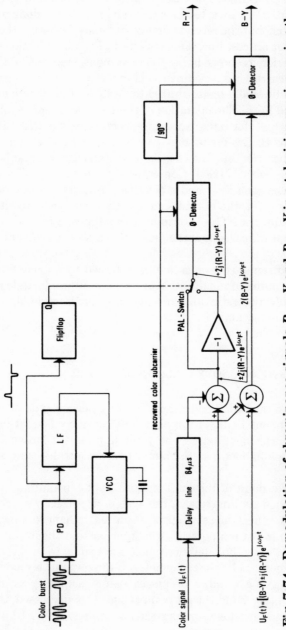
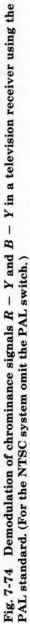

**Fig. 7-74** Demodulation of chrominance signals $R - Y$ and $B - Y$ in a television receiver using the PAL standard. (For the NTSC system omit the PAL switch.)

entiate between phase errors of 0, $2\pi$, $4\pi$, and so on. A phase error of exactly $\pm 2\pi$ corresponds to an extraneous or missed cycle of the system.

In most applications of the PLL it will be of no concern if the system unlocks temporarily from time to time. There are some cases, however, where impulse synchronism is mandatory. If a film is displayed synchronously with a sound recorded on a separate track (tape or cassette), the drives of the film projector and of the tape must operate impulse-synchronously. When the drives are switched on or off, one of them will most certainly lead the other. A phase error of a multiple of $2\pi$ will then build up. To obtain an impulse-synchronous system, a PD must be found which is able to measure phase error over an extended range. We now consider PDs of this kind.

Figure 7-75 shows a phase/frequency detector capable of storing phase errors much greater than $2\pi$. In effect, the range of the phase error is determined by the range of the counter and can be made arbitrarily large. The heart of the circuit is an UP/DOWN counter. It counts 1 upward on every positive-going edge of the signal $u_1$ and 1 downward on every positive-going edge of the signal $u_2$. Most UP/DOWN counters do not have two independent inputs for UP and DOWN counting, but instead one single clock input and a direction (or UP/DOWN) input which determines whether the counter will count up or down. A converter circuit is therefore placed between the $u_1$ and $u_2$ inputs and the counter. When the UP/DOWN counter is a 4-bit binary counter, for example, its content can vary from 0 to 15. At the start of operation the counter would preferably be preset to approximately half-scale, such as to the number 7 (corresponding to a binary pattern of 0111). The most significant bit (MSB) of the counter can then be used as the output signal $\overline{u_d}$ of the PD. Whenever the PLL has become locked, the content of the counter is switched back and forth between the values 7 and 8. The output signal is then simply the averaged duty-cycle ratio of the MSB signal.

The output signal $\overline{u_d}$ is plotted against phase error $\theta_e$ in Fig. 7-76. If the phase error is just $\pi$, the counter switches back and forth between 7 and 8 with a duty cycle of 50 percent. Then $\overline{u_d}$ is half the supply voltage, which is defined as $\overline{u_d} = 0$ (refer to Sec. 4-1-1). If the phase error moves in the positive direction, the duty cycle becomes greater, and $\overline{u_d}$ is positive. If the phase error becomes less than $\pi$, however, the duty cycle is less than 50 percent, and $\overline{u_d}$ goes negative. Now if the phase error exceeds $2\pi$, the counter will toggle between 8 and 9, and if the phase error becomes even greater than $4\pi$, the counter will toggle between 9 and 10, and so on. In any case, for a phase error greater than $2\pi$, the MSB is continuously ON ($\overline{u_d}$ is at its positive extreme). This forces the VCO to increase its frequency until the counter will toggle again at a content between 7 and 8.

Similar things happen when the phase error becomes less than zero.

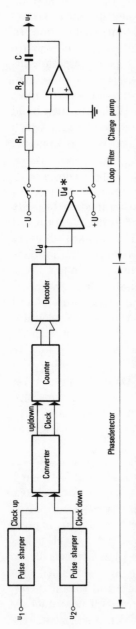

Fig. 7-75   Block diagram of a phase/frequency detector (similar to type 4 in Table 2-1) modified for impulse-synchronous operation. The PD is followed by a charge pump loop filter. ($\overline{u_d}*$ denotes the inverted signal $\overline{u_d}$.)

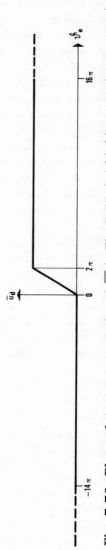

Fig. 7-76   Phase-detector output signal $\overline{u_d}$ in Fig. 7-75. A 4-bit binary UP/DN counter is plotted against phase error for the PD shown assumed.

The counter then first toggles between 6 and 7. The MSB is now fully OFF, and $\overline{u_d}$ is at its negative extreme. If the counting range is 0–15 as assumed previously, a maximum phase error of $16\pi$ and a minimum phase error of $-14\pi$ can be recorded. If this limit is exceeded in either direction, the counter overflows or underflows. If this occurs, 16 cycles get lost or 16 cycles are added erroneously. This phenomenon can be avoided, however, by making the counting range so large that the range of the counter is never exceeded.

In Fig. 7-75 a loop filter is also included. It is most convenient to use a charge pump for the loop filter, as discussed earlier (Table 2-2). A practical phase/frequency detector circuit for impulse-synchronous operation is shown in Fig. 7-77. It is constructed from two cascaded 4-bit binary UP/DOWN counters. The block previously called the converter is shown to the left. The circuit is fabricated entirely from CMOS devices. The range of the counter is 0–255. The most obvious way of operating the counter would be to preset it initially at half-scale, so that it would switch between 127 and 128 in the settled state. The phase-error range would then be from

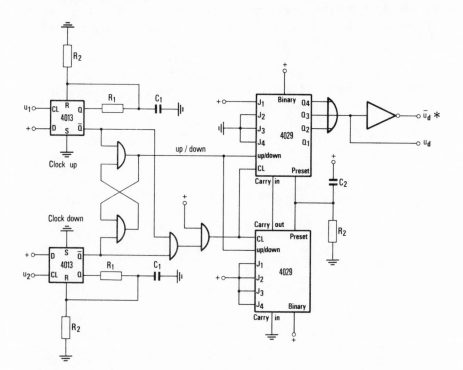

**Fig. 7-77**  **Practical phase/frequency detector circuit for impulse-synchronous operation. ($\overline{u_d}^*$ denotes the inverted signal $\overline{u_d}$.)**

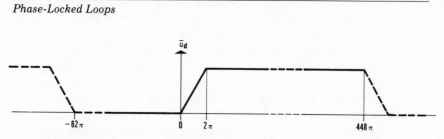

**Fig. 7-78 Phase-detector output signal $\overline{u_d^*}$ plotted against phase error for the PD shown in Fig. 7-77.**

$-127 \cdot 2\pi$ to $+128 \cdot 2\pi$. As shown in Fig. 7-77, this counter could also work with an asymmetric phase-error range. This could be of interest with motor drives where larger positive than negative phase errors are expected. The counter in Fig. 7-77 operates in such a way that it toggles between 31 and 32 in the synchronized state.

The output signal $\overline{u_d}$ of the PD is plotted against phase error in Fig. 7-78 for the asymmetrical case. The output signal is obtained simply by decoding outputs $Q_2$, $Q_3$, and $Q_4$ by means of an OR gate. If the content of the cascade of two counters is equal to or greater than 32, at least one of the outputs $Q_2$, $Q_3$, and $Q_4$ is HIGH. The inverted signal $\overline{u_d^*}$ is generated in addition. (The asterisk is used here to denote logic inversion, the overbar to indicate average value.) Both $\overline{u_d}$ and $\overline{u_d^*}$ signals are used to control the UP and DOWN inputs, respectively, of a charge pump, as shown in Fig. 7-75. The PRESET input in Fig. 7-77 is used to set the counter initially to 32. An auto-start feature is provided by the network $R_2$ and $C_2$.

# 8 MEASURING PLL PARAMETERS

The parameters of a PLL IC or a related circuit are often insufficiently specified in data sheets. It is therefore advantageous if the users are able to measure these parameters themselves. It will be shown in this section that the relevant parameters such as $K_0$, $K_d$, $\omega_n$, $\zeta$, and many others are easily measured with the standard equipment available even in a hobbyist's lab, i.e., an oscilloscope and a waveform generator. For the following measurements a multifunction integrated circuit of type XR-S200 (EXAR) has been arbitrarily selected as the DUT.

## 8-1 MEASUREMENT OF CENTER FREQUENCY $f_0$

Only the VCO portion of the DUT is used for this measurement (Fig. 8-1). The two pins of the symmetrical VCO input are grounded. Consequently the VCO will oscillate at the center frequency $f_0$. The center frequency is now easily measured with an oscilloscope (Fig. 8-2a). If the values $C_{ext} = 82$ nF and $R_0 = 2.2$ k$\Omega$ are chosen for the external components, the VCO oscillates at a center frequency $f_0$ of 6.54 kHz.

## 8-2 MEASUREMENT OF VCO GAIN $K_0$

The same test circuit (Fig. 8-1) can be used to measure the VCO gain $K_0$. One of the VCO inputs (VCO in) stays grounded; a variable dc voltage is applied to the other. This signal corresponds to the loop filter output signal $u_f$. By definition the VCO gain $K_0$ is equal to the variation of VCO angular frequency $\Delta\omega_0$ related to a variation of the $u_f$ signal by $\Delta u_f = 1$ V. As shown in Fig. 8-2b, the VCO frequency falls to 5.78 kHz for $u_f = 1$ V, which corresponds to a variation of 0.76 kHz. The sign of the frequency variation is irrelevant here; it would have been positive if the VCO input pins had been connected with inverted polarity. For the VCO gain $K_0$ we now obtain

$$K_0 = \frac{\Delta\omega_0}{\Delta u_f} = \frac{2\pi \cdot 0.76 \times 10^3}{1} = 4.78 \times 10^3 \text{ s}^{-1}\text{V}^{-1}$$

The test shows furthermore that $K_0$ would be larger by a factor of 10 if $\omega_0$ had been chosen larger by a factor of 10. Hence, for this device, $K_0$

varies in proportion to $\omega_0$. When $R_0 = 2.2$ kΩ is chosen, $K_0$ is given by the general equation

$$K_0 = 0.73f_0 \qquad s^{-1}V^{-1} \tag{8-1}$$

where $f_0$ is in hertz. For example, for $f_0 = 1$ kHz, we find $K_0 = 730$ s$^{-1}$ V$^{-1}$.

### 8-3 MEASUREMENT OF PHASE-DETECTOR GAIN $K_d$

The test circuit of Fig. 8-3 is used for the measurement of $K_d$. The PD of the XR-S200 is shown as an operational multiplier (type 1 according to Table 2-1). In the configuration shown, the PD and the VCO are used. The VCO output signal is coupled capacitively to one of the PD inputs. A variable dc level is applied to the other input.

The measurement procedure is explained by the waveforms in Fig. 8-4. In Fig. 8-4$a$ the signal applied to phase-detector input 1 (PD in # 1) is a sine wave, and the signal applied to phase-detector input 2 (PD in # 2) is a square wave (usually the VCO output signal). It is furthermore assumed that both signals are in phase. Consequently the phase error $\theta_e$ is 90°. The phase-detector output signal (PD out) is a full-wave rectified sine signal; its average value is given by

$$\overline{u_d} = K_d \sin 90° = K_d$$

i.e. the measured output signal is identical with $K_d$.

If a dc voltage is applied to PD in #1, whose level is equal to the average value of the previously applied sine signal, the output signal of the phase detector becomes a square wave having a peak amplitude $K_d$. In most data sheets $K_d$ is specified as a function of the reference signal level; a sine signal is usually chosen for the reference signal, and the signal level is usually given as an rms value. In Fig. 8-4$b$ we see that the rms

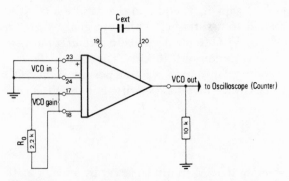

**Fig. 8-1  Test circuit for the measurement of the center frequency $\omega_0$ and the VCO gain $K_0$.**

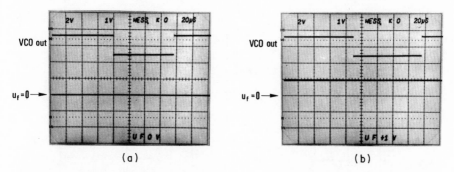

**Fig. 8-2** (*a*) Measurement of $\omega_0$, $u_f = 0$ V. (*b*) Measurement of $K_0$, $u_f = 1$ V.

value of the sine signal is 1.11 times its average value. [For a sine signal of peak amplitude 1, the rms value is $1/\sqrt{2}$, the average value (linear average) is $2/\pi$.] Consequently, to measure $K_d$ for a reference signal level of 1 V rms, we have to apply a dc level of $1/1.11 \approx 0.9$ V to PD in #1 and then simply measure the peak value of the output signal.

The phase-detector gain of the DUT in Fig. 8-3 was measured at four different levels of the reference signal: at 10, 30, 40, and 100 mV rms. The measured signals are displayed in Fig. 8-5. The four oscillograms show the results for $u_1 = 10, 20, 30,$ and 40 mV rms, respectively. As Fig. 8-3 demonstrates, this PD has a symmetrical output. Three signals are displayed

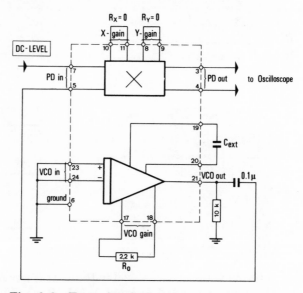

**Fig. 8-3 Test circuit for the measurement of phase-detector gain $K_d$.**

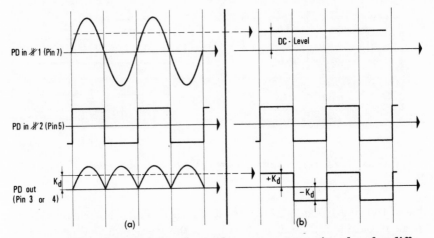

Fig. 8-4 Waveforms of the phase-detector output signal $u_d$ for different signals applied to input 1. (*a*) For a sine-wave signal. (*b*) For a dc level.

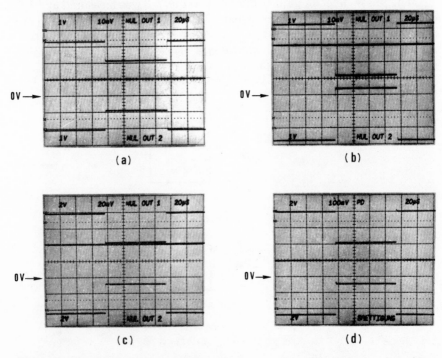

Fig. 8-5 Waveforms measured with the test circuit of Fig. 8-3 for different dc levels applied to input 1. (*a*) 10 mV. (*b*) 30 mV. (*c*) 40 mV. (*d*) 100 mV.

in each oscillogram: the top trace shows multiplier output 1 (MUL OUT 1), the center trace the reference signal (dc), and the bottom trace multiplier output 2 (MUL OUT 2).

Since the output of the PD is a symmetrical signal, the phase-detector gain is *twice* the measured peak amplitude of the square wave.

The measured $K_d$ values are plotted against the reference signal level in Fig. 8-6. It is clearly observed that the PD becomes saturated at signal levels above 40 mV rms.

### 8-4 MEASUREMENT OF HOLD RANGE $\Delta\omega_H$ AND LOCK RANGE $\Delta\omega_L$

To measure these parameters a full PLL circuit must be built. This is easily done by adding a loop filter to the test circuit of Fig. 8-3 and by closing the loop. The new test circuit of Fig. 8-7 is then obtained. A type 1 loop filter (Table 2-2) is chosen for simplicity; it consists of two on-chip resistors, $R_1 = 6$ k$\Omega$ each, and the external capacitor $C$. A signal generator is applied to the reference input (pin 7) of the DUT. Both reference signal $u_1$ and VCO output signal $u_2$ are displayed vs. time on an oscilloscope. The oscilloscope is triggered on $u_2$.

The frequency of the signal generator is now varied manually until lock-in is observed (Fig. 8-8). In the case of Fig. 8-8$a$ the reference frequency $|f_1|$ is far away from the center frequency $f_0$, and the system is unlocked. Fig. 8-8$b$ shows the situation where $f_1$ has come closer to the center frequency. The PLL is trying to pull in the VCO. Frequency modulation of $u_2$ is easily observable. If the reference frequency approaches $f_0$ slightly more, the PLL suddenly locks (Fig. 8-8$c$).

Determination of the hold range $\Delta f_H$ and the lock range $\Delta f_L$ is very simple. The hold range is measured by slowly varying the reference frequency $f_1$ and monitoring the upper and lower values of $f_1$ where the system unlocks. In a similar way the lock range is determined by monitoring the upper and lower values of $f_1$ where the system becomes locked.

The value obtained for the lock range is sometimes higher than the value specified by the manufacturer. In effect, the method shown in Fig. 8-7 yields a value which is between the lock range and the pull-in range. (Refer also to Sec. 3-2.) If the reference frequency $f_1$ is swept very slowly, the measured figure is closer to the pull-in range; if it is swept faster, the measured value comes closer to the lock range.

### 8-5 MEASUREMENT OF NATURAL FREQUENCY $\omega_n$ AND DAMPING FACTOR $\zeta$

The PLL circuit of Fig. 8-7 is used for the following measurements. To measure the natural frequency $\omega_n$ and the damping factor $\zeta$ of a PLL, we

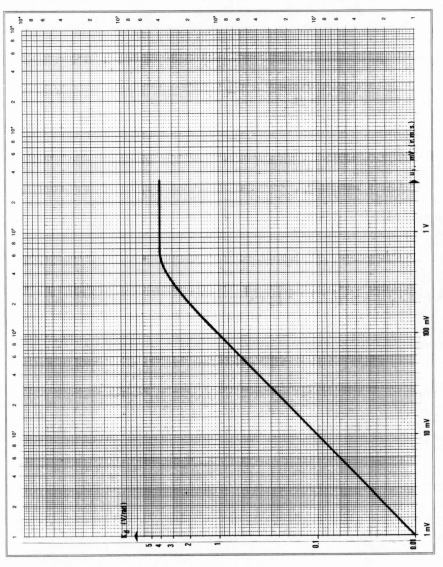

**Fig. 8-6** Plot of $K_d$ against input voltage level in millivolts rms.

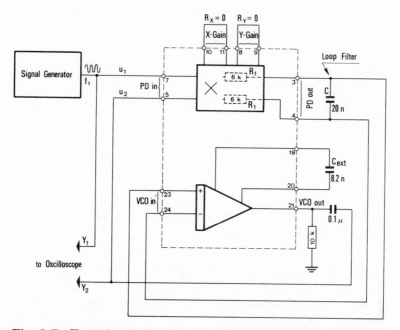

Fig. 8-7 Test circuit for the measurement of hold range $\Delta\omega_H$ and lock range $\Delta\omega_L$.

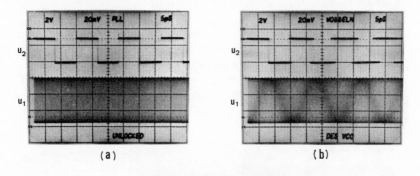

(a)                                (b)

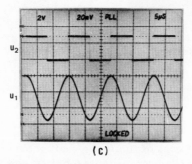

(c)

Fig. 8-8 Waveforms of signals $u_1$ and $u_2$ in the test circuit of Fig. 8-7. (a) PLL unlocked. (b) PLL near locking. (c) PLL locked.

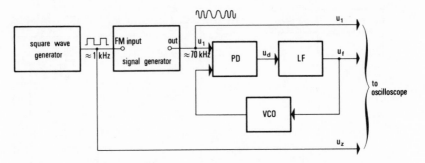

**Fig. 8-9 Test circuit for the measurement of natural frequency $\omega_n$ and damping factor $\zeta$.**

apply a disturbance to the PLL which forces the system to settle at a different stable state. This is most easily done by modulating the reference frequency with a square-wave signal. The corresponding test circuit is shown in Fig. 8-9. The PLL under test operates at a center frequency of approximately 70 kHz. The frequency of the signal generator is modulated by a square-wave generator. Of course, the modulating frequency must be chosen much smaller than the center frequency, for example 1 kHz.

If the frequency of the signal generator is abruptly changed, a phase error $\theta_e$ results. The output signal $u_f$ of the loop filter can be considered a measure of the average phase error. Thus the transient response of the PLL can be easily analyzed by recording $u_f$ on an oscilloscope. This is shown by the waveforms in Fig. 8-10a, which displays the signals $u_1$ (reference signal, top trace), $u_f$ (center trace), and the 1-kHz square wave

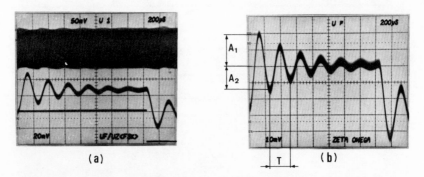

**Fig. 8-10 Waveforms of the test circuit shown in Fig. 8-9. (a) Top trace—input signal $u_1$; center trace—output signal $u_f$ of the loop filter; bottom trace—modulating signal, 1-kHz square wave. (b) Enlarged view of the $u_f$ waveform shown in (a), center trace.**

276

(bottom trace). The oscilloscope is triggered on the 1-kHz square wave. Because the reference signal has a much higher frequency than the 1-kHz square wave and is by no means synchronized with the latter, it is displayed as a bar only.

The $u_f$ signal performs a damped oscillation on every transient of the 1-kHz square wave and settles at a stable level thereafter. Now $\zeta$ and $\omega_n$ can be calculated from the waveform of $u_f$. Figure 8-10$b$ is an enlarged view of this signal; $\zeta$ can be calculated from the ratio of the amplitudes of two subsequent half-waves $A_1$ and $A_2$. (Any pair of subsequent half-waves may be chosen.) The damping factor is given by

$$\zeta = -\frac{\ln (A_1/A_2)}{(\pi^2 + [\ln (A_1/A_2)]^2)^{1/2}}$$

The natural frequency $\omega_n$ is calculated from the period $T$ of one oscillation in Fig. 8-10$b$ according to

$$\omega_n = \frac{2\pi}{T\sqrt{1 - \zeta^2}}$$

Let us now evaluate numerically the waveform of Fig. 8-10$b$. For $A_1$ and $A_2$ we read approximately 1.9 and 1.5 divisions, respectively. Consequently we obtain

$$\zeta \approx 0.08$$

For $T$ we find $T \approx 240$ $\mu$s. Hence $\omega_n$ is

$$\omega_n \approx 26.0 \times 10^3 \text{ s}^{-1}$$

or

$$f_n \approx 4.1 \text{ kHz}$$

To complete this measurement, let us calculate the values of $\omega_n$ and $\zeta$ using the theory of the linear PLL [Eq. (3-6)] and check whether our measurements agree with the predicted results. Using Eq. (8-1), we obtain for the VCO gain

$$K_0 = 0.73 \cdot 70 \times 10^3 = 51.1 \times 10^3 \text{ s}^{-1}\text{V}^{-1}$$

According to Fig. 8-10$a$ the amplitude of the reference signal is about 120 mV peak-to-peak, which corresponds to an rms value of 43 mV. For this signal level we read a phase-detector gain $K_d \approx 3.7$ V from Fig. 8-6. The two time constants $\tau_1$ and $\tau_2$ can be derived from the test circuit in Fig. 8-7,

$$\tau_1 = 2.6 \text{ k}\Omega \cdot 20 \text{ nF} = 240 \text{ }\mu\text{s}$$

$$\tau_2 = 0$$

Using Eq. (3-6) we finally get

$$\omega_n = \left(\frac{K_0 K_d}{\tau_1}\right)^{1/2} = \left(\frac{51.1 \times 10^3 \cdot 3.7}{240 \times 10^{-6}}\right)^{1/2} = 28 \times 10^3 \text{ s}^{-1}$$

$$\zeta = \frac{1}{2}\,\omega_n\left(\tau_2 + \frac{1}{K_0 K_d}\right) = \frac{28 \times 10^3}{2 \cdot 51.1 \times 10^3 \cdot 3.7} = 0.07$$

This agrees well with the experimental measurements.

Note that this experimental method of measuring $\omega_n$ and $\zeta$ is applicable only for $\zeta < 1$. This requirement is met, however, in most cases. If $\zeta$ were greater than 1, the transient response would become aperiodic, and it would become impossible to define the values $A_1$ and $A_2$ in Fig. 8-10a. In the example of Fig. 8-10a the damping factor $\zeta$ has purposely been chosen too small in order to get a marked oscillatory transient. An underdamped system is often obtained when a loop filter without a zero (type 1 or 3 according to Table 2-1) is chosen. To increase $\zeta$, a type 2 filter should be specified.

### 8-6 MEASUREMENT OF THE PHASE-TRANSFER FUNCTION $H(j\omega)$ AND THE 3-dB BANDWIDTH $\omega_{3dB}$

As discussed in Sec. 7-1, the dynamic performance of the tracking filter is best described by the phase-transfer function $H(j\omega)$. If this transfer function has been measured, the 3-dB bandwidth $\omega_{3dB}$ is obtained automatically. There are different methods of measuring $H(j\omega)$; some of these use PM techniques,[49] others use frequency modulation.

The PM technique has the advantage that the maximum phase error $\theta_e$ never exceeds the peak phase deviation $\Delta\phi$ [refer to Eq. (7-22)]. If $\Delta\phi$ is restricted to small values (such as $< \pi/4$), the PD will never operate in its nonlinear region. Unfortunately, most signal generators used in the lab can be frequency-modulated but not phase-modulated.

To enable the measurement of $H(j\omega)$ by off-the-shelf instruments, we will consider the FM techniques in more detail. When using the FM method, we must be careful that the peak phase error $\hat{\theta}_e$ does not exceed a practical limit of say $\pi/4$. As we know from Fig. 3-15, the phase error can become very large if the modulating frequency $\omega_m$ approaches the natural frequency $\omega_n$ of the PLL. Furthermore, the peak phase error is proportional to the peak frequency deviation $\Delta\omega$. If $\hat{\theta}_e$ is required not to exceed, say, $\pi/4$, we see from Fig. 3-15 that $\Delta\omega$ should be limited to less than $\omega_n/2$ if a damping factor $\zeta$ of 0.3 is assumed.

In Sec. 8-5 $\omega_n$ has been determined as $\omega_n \approx 26.0 \times 10^3 \text{ s}^{-1}$ or $f_n \approx 4.1$ kHz. We should therefore choose a peak frequency deviation $\Delta f$ of about 2 kHz or less. But even without knowing the value of $\omega_n$ there is a simple

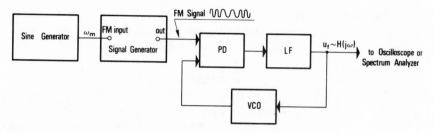

**Fig. 8-11   Test circuit for the measurement of $H(j\omega)$.**

way of experimentally determining the maximum value of $\Delta f$. The test setup for the measurement of $H(j\omega)$ is shown in Fig. 8-11. According to Eq. (7-8) the amplitude of the signal $u_f$ is proportional to $|H(j\omega)|$. If the modulating signal in this test circuit is a sine wave, $u_f$ must also be a sine wave, at least as long as the PD is operating in its linear region. To check whether the initially chosen peak frequency deviation $\Delta\omega$ is adequate, we monitor the $u_f$ waveform on an oscilloscope and manually sweep the modulating frequency $f_m$ from 0 to, say, 10 kHz. If $f_m$ comes close to the natural frequency $f_n$, a resonance peak is observed on the scope. If the $u_f$ waveform stays undistorted at its peak amplitude, the peak frequency deviation $\Delta f$ is adequate (refer to Fig. 8-12a).

If too large a peak frequency deviation has been selected, the $u_f$ waveforms becomes distorted, as shown in Fig. 8-12b. This indicates that $\hat{\theta}_e$ has become so large that the phase detector now operates near saturation. If such a distorted waveform is observed, the peak frequency deviation simply must be reduced.

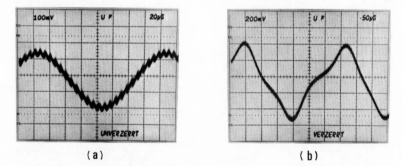

(a)                                        (b)

**Fig. 8-12   Output signal of the loop filter $u_f$ measured by the test circuit of Fig. 8-11 (a) The peak frequency deviation of the FM modulation is chosen low enough to enable linear operation of the PD. (b) The peak frequency deviation is too large. The PD is operating in its nonlinear region.**

279

The phase-transfer function can now be measured by plotting the amplitude of the $u_f$ signal against the modulating frequency $\omega_m$. This can be done manually, but the measurement can also be automated by means of a spectrum analyzer.

In this case the modulating frequency $\omega_m$ must be generated by a sweep generator. If the phase-transfer function $H(j\omega)$ has to be recorded within a range of modulating frequencies of, say, 0 to 10 kHz, a sweep range of 10 kHz has to be specified for both the spectrum analyzer and the sweep generator. However, the sweep rate has to be chosen differently for each instrument.

The sweep rate of the spectrum analyzer depends on the specified spectral resolution. In this example a spectral resolution of 300 Hz has been chosen. The spectrum analyzer then scans the frequency spectrum from 0 to 10 kHz once every second (Fig. 8-13). To avoid intermodulation between spectrum analyzer and sweep generator, the modulating frequency $f_m$ should stay quasistationary during one sweep period of the spectrum analyzer. Consequently, the chosen sweep rate of the sweep generator should be much slower than the sweep rate of the spectrum analyzer. In our example the sweep generator scans through the frequency range from 0 to 10 kHz within 100 s, as shown in Fig. 8-13.

The phase-transfer function $H(j\omega)$ recorded with this technique is shown in Fig. 8-14. The spectrum was recorded by means of a Polaroid camera; because the spectrum of 0–10 kHz is scanned in a time interval of 100 s, the shutter of the camera had to remain open for 100 s. As an alternative, the spectrum could have been recorded first by a storage oscilloscope and photographed afterward with a normal exposure time.

The phase-transfer function $H(j\omega)$ of the PLL is shown in Fig. 8-14 for two different conditions. Fig. 8-14$a$ is the "correct" measurement; the

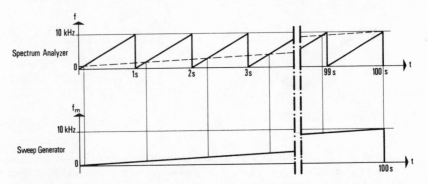

**Fig. 8-13 Measurement of phase-transfer function $H(j\omega)$. The waveforms explain the timing of the spectrum analyzer and the sweep generator.**

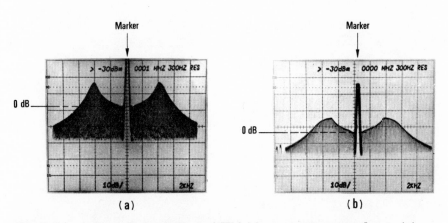

**Fig. 8-14** **Automatic recording of** $H(j\omega)$ **by a spectrum analyzer.** (*a*) **The peak frequency deviation was chosen so that the PD operates in its linear region** (*b*) **The peak frequency deviation is too high, and the transfer function is distorted.**

peak frequency deviation was chosen small enough to avoid nonlinear operation of the PD. The frequency scale is symmetrical with respect to the marker, which indicates the origin of the frequency scale. The left half of the spectrum covers the range of negative frequencies from 0 to $-10$ kHz and can be discarded here.

A low damping factor of $\zeta = 0.08$ was purposely chosen for this measurement. A large resonance peak is therefore observed at the natural frequency of $f_n \approx 4.1$ kHz. Note that this value was measured by a different method in Sec. 8-5.

The 3-dB frequency $f_{3dB}$ is by definition the frequency for which $|H(j\omega)|$ is 3 dB lower than the dc value $|H(j0)|$. From Fig. 8-14*a* we obtain

$$f_{3dB} \approx 6.8 \text{ kHz}$$

Fig. 8-14*b* shows the same measurement, but the peak frequency deviation $\Delta\omega$ was purposely chosen so high as to cause nonlinear operation of the PD. The measurement is dramatically corrupted by this procedure; the amplitude of the spectrum is heavily compressed, and the curve shows a discontinuity. Nevertheless locations of the natural frequency $f_n$ and the 3-dB frequency $f_{3dB}$ were not shifted to any considerable extent.

## Appendix A
## THE PULL-IN PROCESS[*]

The pull-in process can be calculated to an approximation only. Different authors have developed approaches which differ from each other slightly.[2,50] We will present here an approximation found by the author. Our aim is to find (1) the pull-in range $\Delta\omega_P$, i.e., the maximum frequency offset $\omega_1 - \omega_0$ at which a pull-in process is still possible, and (2) the pull-in time $T_P$, i.e., the time required for the PLL to get locked. Our calculation is based on a second-order linear PLL having a type 2 or type 3 loop filter (Table 2-1).

It is assumed that the power supply of the PLL is switched on at time $t = 0$, and that the initial frequency offset $\omega_1 - \omega_0$ is greater than the lock range $\Delta\omega_L$. Therefore, the PLL will not lock immediately, and the VCO will oscillate initially at the center frequency $\omega_0$ (Fig. A-1, bottom trace). The instantaneous frequency of the VCO is denoted by $\omega_2$. The output signal $u_d$ of the PD is a periodic signal; its angular frequency is given by the difference $\omega_1 - \omega_0$,

$$u_d(t) = K_d \sin [(\omega_1 - \omega_0)t + \phi] \tag{A-1}$$

where $\phi$ is the zero phase.

We must make some simplifications for the following calculation: Assume first that the difference $\omega_1 - \omega_0$ is much greater than both corner frequencies of the loop filter [$1/(\tau_1 + \tau_2)$ and $1/\tau_2$]. Then the gain of the loop filter can be approximated by a constant $F_H$ which is for a type 2 filter,

$$F_H \approx \frac{\tau_2}{\tau_1 + \tau_2} \tag{A-2a}$$

and for a type 3 filter,

$$F_H \approx \frac{\tau_2}{\tau_1} \tag{A-2b}$$

Because $\tau_1 \gg \tau_2$ in most cases, we can simply write

$$F_H \approx \frac{\tau_2}{\tau_1} \tag{A-3}$$

for both types of loop filters. The output signal $u_f$ of the loop filter is then given by

$$u_f(t) = F_H u_d(t) = F_H K_d \sin [(\omega_1 - \omega_0)t + \phi] \tag{A-4}$$

---

[*]Appendix to Sec. 3-2-3.

This signal causes a frequency modulation of the VCO output signal (Fig. A-1, bottom curve); its peak frequency deviation is given by $F_H K_0 K_d$. A closer look at this illustration reveals that the difference $\omega_1 - \omega_2$ between reference and output frequencies is no longer a constant, but varies with the frequency modulation of the VCO signal. Figure A-1 shows that the difference $\omega_1 - \omega_2$ becomes smaller during the time intervals when the frequency of the VCO is increased, and vice versa. The signal $u_d(t)$ becomes therefore inharmonic (see top curve in Fig. A-1), i.e., the duration of the positive half-waves is longer than that of the negative ones. Consequently the average value $\overline{u_d}$ of the phase-detector output signal is *not zero*, but slightly *positive*. This slowly pulls the frequency $\omega_2$ of the VCO in the positive direction.

Under certain conditions to be discussed in the following, this process is *regenerative*, i.e., the VCO is pulled to a frequency close enough to $\omega_1$ that the PLL finally locks. For the calculation of the pull-in process we assume that the frequency $\omega_2$ of the VCO has already been pulled somewhat in the direction of $\omega_1$ (Fig. A-2). The average angular frequency of the VCO is denoted by $\omega_{20}$, and the average frequency offset is $\Delta\omega = \omega_1 - \omega_{20}$. Next the average value $\overline{u_d}$ is calculated as a function of the frequency offset $\Delta\omega$. For an exact solution, the waveform of the signal $u_d(t)$ should be known. This calculation is extremely difficult, however, so we try to get an approximation.

Four distinct points $A$, $B$, $C$ and $D$ of the signal $\omega_2(t)$ are known exactly. Hence, we can derive a simplified expression for its waveforms as well as for the

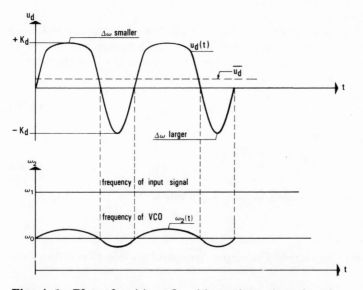

**Fig. A-1 Plot of $u_d(t)$ and $\omega_2(t)$ against time for the unlocked PLL. It is assumed that the power supply of the PLL has been turned on at $t = 0$.**

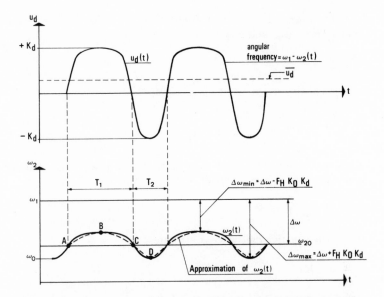

Fig. A-2 Plot of $u_d(t)$ and $\omega_2(t)$ for the unlocked PLL during the pull-in process. It is assumed that the frequency of the VCO has already been pulled somewhat toward the reference frequency $\omega_1$.

waveform $u_d(t)$, which is proportional to $\omega_2(t)$. At points $A$ and $C$ the instantaneous frequency $\omega_2(t)$ is exactly $\omega_{20}$. At point $B$, $\omega_2(t)$ is at its positive peak deviation, that is, $\omega_2(t) = \omega_{20} + F_H K_0 K_d$. By analogy the frequency at point $D$ is $\omega_2(t) = \omega_{20} - F_H K_0 K_d$. The function $\omega_2(t)$ is now approximated by two sine halfwaves (see the dashed curve in Fig. A-2). This simplification allows us now to calculate the average frequencies $\overline{\omega_{2+}}$ and $\overline{\omega_{2-}}$ during the positive and negative half-waves, respectively. The average value of a half-wave is obtained by multiplying its peak amplitude by $2/\pi$. For $\overline{\omega_{2+}}$ and $\overline{\omega_{2-}}$ we get

$$\overline{\omega_{2+}} = \omega_{20} + \frac{2}{\pi} F_H K_0 K_d \tag{A-5a}$$

$$\overline{\omega_{2-}} = \omega_{20} - \frac{2}{\pi} F_H K_0 K_d \tag{A-5b}$$

The average values of the frequency offset $\Delta\omega(t) = \omega_1 - \omega_{20}(t)$ in the positive and negative half-waves can now be calculated as well:

$$\overline{\Delta\omega_+} = \Delta\omega - \frac{2}{\pi} F_H K_0 K_d \tag{A-6a}$$

$$\overline{\Delta\omega_-} = \Delta\omega + \frac{2}{\pi} F_H K_0 K_d \tag{A-6b}$$

The duration of the half-waves $T_1$ and $T_2$ (Fig. A-2) can then be approximated from $\overline{\Delta\omega_{2+}}$ and $\overline{\Delta\omega_{2-}}$:

$$T_1 = \frac{1}{2}\frac{2\pi}{\Delta\omega_+} = \frac{\pi}{\Delta\omega - \dfrac{2}{\pi}F_H K_d K_0} \tag{A-7a}$$

$$T_2 = \frac{1}{2}\frac{2\pi}{\Delta\omega_-} = \frac{\pi}{\Delta\omega + \dfrac{2}{\pi}F_H K_d K_0} \tag{A-7b}$$

Knowing $T_1$ and $T_2$, we can now calculate $\overline{u_d}$. The positive half-wave of $u_d(t)$ contributes an average value of

$$\frac{2}{\pi}K_d\frac{T_1}{T_1 + T_2}$$

to $\overline{u_d}$; the negative half-wave contributes

$$-\frac{2}{\pi}K_d\frac{T_2}{T_1 + T_2}$$

Hence $\overline{u_d}$ is given by

$$\overline{u_d} = \frac{2}{\pi}K_d\frac{T_1 - T_2}{T_1 + T_2} \tag{A-8}$$

To get a simple expression for $\overline{u_d}$, $T_1$ and $T_2$ are expanded to a Taylor series and terms of second and higher order are neglected. Then we get for $T_1$ and $T_2$:

$$T_1 = \frac{\pi}{\Delta\omega[1 - (2/\pi)F_H K_d K_0/\Delta\omega]} \approx \frac{\pi}{\Delta\omega}\left(1 + \frac{2}{\pi}\frac{F_H K_d K_0}{\Delta\omega}\right) \tag{A-9a}$$

$$T_2 = \frac{\pi}{\Delta\omega[1 + (2/\pi)F_H K_d K_0/\Delta\omega]} \approx \frac{\pi}{\Delta\omega}\left(1 - \frac{2}{\pi}\frac{F_H K_d K_0}{\Delta\omega}\right) \tag{A-9b}$$

With these approximations $\overline{u_d}$ becomes

$$\boxed{\overline{u_d} = \frac{8F_H K_d^2 K_0}{\pi^2\,\Delta\omega}} \tag{A-10}$$

Thus $\overline{u_d}$ is inversely proportional to $\Delta\omega$. This is valid for large values of $\Delta\omega$ only, because $\overline{u_d}$ can never become greater than $K_d$ [refer to Eq. (2-3) or Fig. A-2]. In Fig. A-3 $\overline{u_d}$ is plotted against $\Delta\omega$ (solid curve); the curve is a hyperbola for large values of $\Delta\omega$, but approaches $K_d$ for small $\Delta\omega$ values.

Now we can draw a simple mathematical model of the pull-in process (Fig. A-4). The transfer characteristics of the blocks are shown graphically. The transfer function for the VCO is

$$\omega_{20} = \omega_0 + K_0\overline{u_f} \tag{A-11}$$

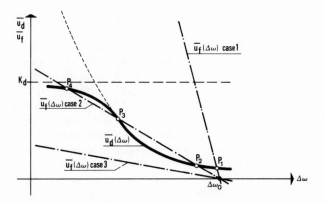

**Fig. A-3 Plot of the average signals $\overline{u_d}(t)$ and $\overline{u_f}(t)$ against frequency offset $\Delta\omega$ for the unlocked PLL.**

With the abbreviations $\Delta\omega = \omega_1 - \omega_{20}$ and $\Delta\omega_0 = \omega_1 - \omega_0$ ( = initial offset), this can be written as

$$\Delta\omega = \Delta\omega_0 - K_0\overline{u_f}$$

or

$$\overline{u_f} = \frac{\Delta\omega_0 - \Delta\omega}{K_0} \tag{A-12}$$

Plotting $\overline{u_f}$ against $\Delta\omega$ yields a straight line. This line is also shown in Fig. A-3.

From this illustration we can determine whether or not a pull-in process will

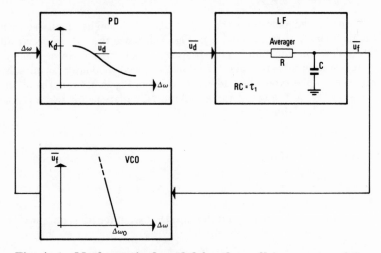

**Fig. A-4 Mathematical model for the pull-in process of the PLL.**

take place. Depending on the slope of $\overline{u}_f$ against $\Delta\omega$, the line $\overline{u}_f(\Delta\omega)$ can intersect with the curve $\overline{u}_d(\Delta\omega)$ at one point (case 1), at three points (case 2), or at no point at all (case 3). Consider case 1 first. The curves $\overline{u}_d(\Delta\omega)$ and $\overline{u}_f(\Delta\omega)$ intersect at point $P_1$. In this case the frequency of the VCO is pulled up slightly, but the system remains "hung" in point $P_1$. An analysis of stability shows that point $P_1$ is a stable point; thus no pull-in process will occur.

In case 2 the curves $\overline{u}_d(\Delta\omega)$ and $\overline{u}_f(\Delta\omega)$ intersect at points $P_2$, $P_3$, and $P_4$. A stability analysis shows that points $P_2$ and $P_4$ are stable, but $P_3$ is unstable. After power-on, the system will thus remain hung at point $P_2$.

Apparently a pull-in process takes place only when there is no point of intersection, as in case 3. The two curves do not intersect if the equation

$$\overline{u}_f(\Delta\omega) = \overline{u}_d(\Delta\omega) \tag{A-13}$$

does not have a real solution. Equation (A-13) is quadratic. Its solutions are

$$\Delta\omega_{1,2} = \frac{\Delta\omega_0}{2} \pm \frac{\sqrt{\Delta\omega_0^2 - (32/\pi^2)K_0^2 K_d^2 F_H}}{2} \tag{A-14}$$

The roots become complex if the discriminant (the expression under the radical) becomes negative. Putting the discriminant equal to zero, yields the limiting case $\Delta\omega_0 = \Delta\omega_P$; that is, the pull-in range $\Delta\omega_P$ is the initial frequency offset $\Delta\omega_0$ for which the discriminant becomes zero. If we perform this calculation, we obtain

$$\Delta\omega_P \approx \frac{4\sqrt{2}}{\pi}\sqrt{2\zeta\omega_n K_0 K_d - \omega_n^2} \tag{A-15}$$

If the PLL is a high-gain loop, this can be simplified to

$$\Delta\omega_P \approx \frac{8}{\pi}\sqrt{\zeta\omega_n K_0 K_d} \tag{A-16}$$

Now we will determine the pull-in time $T_P$; it can be calculated by means of the mathematical model shown in Fig. A-4. Since the pull-in process is a relatively slow phenomenon, higher-frequency components can be discarded. It is therefore acceptable to replace the type 2 or type 3 loop filter (as initially chosen) with the simpler type 1 loop filter, as shown in Fig. A-4. From the mathematical model of Fig. A-4, a set of three differential equations for the variables $\overline{u}_d$, $\overline{u}_f$, and $\Delta\omega$ can be defined:

$$\overline{u}_d = \overline{u}_f + \tau_1 \dot{\overline{u}}_f$$
$$\Delta\omega = \Delta\omega_0 - K_0\overline{u}_f \tag{A-17}$$
$$\overline{u}_d = \frac{8F_H K_d K_0}{\pi^2 \Delta\omega}$$

If $\overline{u}_d$ and $\overline{u}_f$ are eliminated, a differential equation for $\Delta\omega$ is obtained:

$$\frac{8K_d^2 K_0^2 F_H}{\pi^2}\frac{1}{\Delta\omega} + \Delta\omega + \tau_1\dot{\Delta\omega} = \Delta\omega_0 \tag{A-18}$$

This is a nonlinear differential equation, but it can be solved by separation. Separating the terms containing $\Delta\omega$ and $t$, we obtain

$$\tau_1 \frac{\Delta\omega \, d \, \Delta\omega}{-\Delta\omega^2 + \Delta\omega \, \Delta\omega_0 - 8F_H K_d^2 K_0^2/\pi^2} = dt \qquad (A\text{-}19a)$$

After integration on both sides we have

$$\int_{\Delta\omega_0}^{\Delta\omega_L} \frac{\Delta\omega}{-\Delta\omega^2 + \Delta\omega \, \Delta\omega_0 - \dfrac{8F_H K_d^2 K_0^2}{\pi^2}} \, d\Delta\omega = \frac{t}{\tau_1} \bigg|_0^{T_P} = \frac{T_P}{\tau_1} \qquad (A\text{-}19b)$$

The limits of integration on the left-hand side of Eq. (A-19$b$) are $\Delta\omega_0$ and $\Delta_{\omega L}$. At time $t = 0$ the frequency offset is, by definition, $\omega_1 - \omega_0 = \Delta\omega_0$. Hence the lower limit of integration is $\Delta\omega_0$. The pull-in process is terminated when the frequency offset $\Delta\omega$ has reached the value $\Delta\omega_L$ (lock range). When the offset $\Delta\omega$ has been reduced to $\Delta\omega_L$, the system locks rapidly, as was shown in Sec. 3-2-2.

The limits of integration on the right-hand side are zero and $T_P$, since the pull-in process starts at $t = 0$ and is terminated at $t = T_P$. The integral on the left-hand side of Eq. (A-19$b$) is of the type

$$\int \frac{mx + n}{x^2 + 2ax + b} \, dx$$

Its solution is given by

$$\int \frac{mx + n}{x^2 + 2ax + b} \, dx = \frac{m}{2} \ln (x^2 + 2ax + b) + \frac{n - ma}{c} \tan^{-1} \frac{x + a}{c} \qquad (A\text{-}20)$$

Hence there is an explicit solution for the left-hand side in Eq. (A-19$b$). It shows that the argument of the ln function is near 1, so this term can be discarded. Furthermore, the argument of the $\tan^{-1}$ function is very small, so the $\tan^{-1}$ function can be replaced by its argument ($\tan^{-1} x \approx x$). Finally $\Delta\omega_0$ is much greater than $\Delta\omega_L$. Using these approximations and the substitutions in Eq. (3-6), we obtain for $T_P$

$$T_P \approx \frac{\Delta\omega_0^2}{2\zeta\omega_n^3} \qquad (A\text{-}21)$$

## Appendix B
## TRACKING PERFORMANCE OF LINEAR SECOND-ORDER
## LOOPS AT PHASE MODULATION*

In the following discussion, we will calculate the transient response of a second-order linear PLL to a phase step applied to its reference input. The solution presented here is valid for both high-gain and low-gain loops. If a phase step of amplitude $\Delta\Phi$ is applied to the reference input at time $t = 0$, the Laplace transform of the reference phase $\Theta_1(s)$ is given by [refer to Eq. (3-45)]

$$\Theta_1(s) = \frac{\Delta\Phi}{s} \tag{B-1}$$

The Laplace transform of the phase error $\Theta_e(s)$ is given by

$$\Theta_e(s) = \Theta_1(s)H_e(s) \tag{B-2}$$

Assume first that a passive loop filter is used (type 1 or type 2 according to Table 2-1). The error-transfer function $H_e(s)$ [refer to Eq. (3-4)] is then given by

$$H_e(s) = \frac{s^2 + (\omega_n^2/K_0K_d)s}{s^2 + 2s\zeta\omega_n + \omega_n^2} \tag{B-3}$$

Substituting Eq. (B-3) into Eq. (B-2) and applying the inverse Laplace transform (refer also to App. F), we get the following expressions for the phase error $\theta_e(t)$. When $\zeta < 1$,

$$\frac{\theta_e(t)}{\Delta\Phi} = \left[ \cos(\sqrt{1 - \zeta^2}\,\omega_n t) \right.$$
$$\left. + \frac{(\omega_n/K_0K_d) - \zeta}{\sqrt{1 - \zeta^2}} \sin(\sqrt{1 - \zeta^2}\,\omega_n t) \right] \exp(-\zeta\omega_n t) \tag{B-4a}$$

when $\zeta = 1$,

$$\frac{\theta_e(t)}{\Delta\Phi} = \left[ 1 + \left( \frac{\omega_n}{K_0K_d} - 1 \right)\omega_n t \right] \exp(-\omega_n t) \tag{B-4b}$$

and when $\zeta > 1$,

$$\frac{\theta_e(t)}{\Delta\Phi} = \left[ \cosh(\sqrt{\zeta^2 - 1}\,\omega_n t) \right.$$
$$\left. + \frac{(\omega_n/K_0K_d) - \zeta}{\sqrt{\zeta^2 - 1}} \sinh(\sqrt{\zeta^2 - 1}\,\omega_n t) \right] \exp(-\zeta\omega_n t) \tag{B-4c}$$

*Appendix to Sec. 3-3-1.

If the PLL used is a high-gain loop, $\omega_n \ll K_0 K_d$, and the expressions in Eqs. (B-4) can be simplified to read, when $\zeta < 1$,

$$\frac{\theta_e(t)}{\Delta\Phi} = \left[ \cos(\sqrt{1 - \zeta^2}\, \omega_n t) \right.$$
$$\left. - \frac{\zeta}{\sqrt{1 - \zeta^2}} \sin(\sqrt{1 - \zeta^2}\, \omega_n t) \right] \exp(-\zeta\omega_n t) \qquad \text{(B-5a)}$$

when $\zeta = 1$,

$$\frac{\theta_e(t)}{\Delta\Phi} = (1 - \omega_n t) \exp(-\omega_n t) \qquad \text{(B-5b)}$$

and when $\zeta > 1$,

$$\frac{\theta_e(t)}{\Delta\Phi} = \left[ \cosh(\sqrt{\zeta^2 - 1}\, \omega_n t) \right.$$
$$\left. - \frac{\zeta}{\sqrt{\zeta^2 - 1}} \sinh(\sqrt{\zeta^2 - 1}\, \omega_n t) \right] \exp(-\zeta\omega_n t) \qquad \text{(B-5c)}$$

If an active loop filter is used, the error-transfer function is given by [refer to Eq. (3-2)]

$$H_e(s) = \frac{s^2}{s^2 + 2s\zeta\omega_n + \omega_n^2} \qquad \text{(B-6)}$$

Substituting Eq. (B-6) into Eq. (B-2) yields the same expressions for $\theta_e(t)$ as obtained for the passive high-gain PLL [Eqs. (B-5)]. Figure 3-14 is a plot of Eqs. (B-5).

## Appendix C
### *TRACKING PERFORMANCE OF LINEAR SECOND-ORDER LOOPS AT FSK MODULATION**

If a frequency step of size $\Delta\omega$ is applied to the reference input of a PLL at time $t = 0$, the Laplace transform of the reference phase is given by [refer also to Eq. (3-54)]

$$\Theta_1(s) = \frac{\Delta\omega}{s^2} \tag{C-1}$$

For the Laplace transform of the phase error we then have [refer also to Eq. (3-55)]

$$\Theta_e(s) = \Theta_1(s)H_e(s) \tag{C-2}$$

Assume first that a passive loop filter is used. Its transfer function is given by

$$H_e(s) = \frac{s^2 + (\omega_n^2/K_0K_d)s}{s^2 + 2s\zeta\omega_n + \omega_n^2} \tag{C-3}$$

Substituting Eq. (C-3) into Eq. (C-2) and performing the inverse Laplace transform (see also App. F), we get for $\theta_e(t)$, when $\zeta < 1$,

$$\frac{\theta_e(t)}{\Delta\omega/\omega_n} = \frac{\omega_n}{K_0K_d}\left\{1 - \frac{\exp(-\zeta\omega_n t)}{\sqrt{1 - \zeta^2}}\left[\left(\zeta - \frac{K_0K_d}{\omega_n}\right)\sin(\sqrt{1 - \zeta^2}\,\omega_n t)\right.\right.$$
$$\left.\left. + \sqrt{1 - \zeta^2}\cos(\sqrt{1 - \zeta^2}\,\omega_n t)\right]\right\} \tag{C-4a}$$

when $\zeta = 1$,

$$\frac{\theta_e(t)}{\Delta\omega/\omega_n} = \frac{\omega_n}{K_0K_d}(1 - \exp(-\omega_n t)) \tag{C-4b}$$

and when $\zeta > 1$,

$$\frac{\theta_e(t)}{\Delta\omega/\omega_n} = \frac{\omega_n}{K_0K_d}\left\{1 - \frac{\exp(-\zeta\omega_n t)}{\sqrt{\zeta^2 - 1}}\left[\left(\zeta - \frac{K_0K_d}{\omega_n}\right)\sinh(\sqrt{\zeta^2 - 1}\,\omega_n t)\right.\right.$$
$$\left.\left. + \sqrt{\zeta^2 - 1}\cosh(\sqrt{\zeta^2 - 1}\,\omega_n t)\right]\right\} \tag{C-4c}$$

*Appendix to Sec. 3-3-2-2.

In the case of a high-gain loop, $\omega_n/K_0K_d$ is much less than 1, and when $\zeta < 1$ Eq. (C-4) reduces to

$$\frac{\theta_e(t)}{\Delta\omega/\omega_n} = \left[\frac{1}{\sqrt{1-\zeta^2}}\sin(\sqrt{1-\zeta^2}\,\omega_n t)\right]\exp(-\zeta\omega_n t) \qquad \text{(C-5a)}$$

when $\zeta = 1$,

$$\frac{\theta_e(t)}{\Delta\omega/\omega_n} = \omega_n t \cdot \exp(-\omega_n t) \qquad \text{(C-5b)}$$

and when $\zeta > 1$,

$$\frac{\theta_e(t)}{\Delta\omega/\omega_n} = \left[\frac{1}{\sqrt{\zeta^2-1}}\sinh(\sqrt{\zeta^2-1}\,\omega_n t)\right]\exp(-\zeta w_n t) \qquad \text{(C-5c)}$$

If an active loop filter is used, Eqs. (C-5) are valid for any value of loop gain. Figure 3-16 is a plot of Eq. (C-5). Here the phase error $\theta_e(t)$ approaches zero for large $t$. For low-gain loops having a passive loop filter the curves look similar, but the phase error does not decay to zero, it remains finite instead,

$$\theta_e(\infty) = \frac{\Delta\omega}{K_0K_d}$$

[refer also to Eq. (3-58)].

## Appendix D
## THE PULL-IN PROCESS OF THE DIGITAL PLL USING A TYPE 4 PHASE DETECTOR*

The lock-in process of the PLL using the type 4 PD (Table 2-1) was described in Sec. 4-1-2. It was stated there [refer to Eq. (4-4)] that in the unlocked state of the PLL the 4 PD generates an output signal which is proportional to the frequency offset $\Delta\omega = \omega_1 - \omega_2$, where $\omega_1$ is the reference (input) frequency and $\omega_2$ the frequency of the VCO.

In this appendix we will analyze in more detail the performance of the PD in the unlocked state of the PLL. In particular, we will present a mathematical derivation of Eqs. (4-4) and (4-5). The DPLL to be discussed is shown in Fig. 4-4b. For simplicity we assume that $N = 1$ and furthermore that the PLL is initially unlocked and that $\omega_1 > \omega_2$. As shown by the waveforms in Fig. 4-3c, then only the UP output of the PD is pulsed, while the DOWN output stays HIGH (inactive) all the time. We remember that the UP output is pulsed LOW (active) by the negative-going edge of the signal $u_1$ and HIGH by the negative-going edge of the signal $u_2$.

Assume now that the signal $u_1$ performed a negative transition at time $t = 0$, as shown in curve $a$ of Fig. D-1. This transition pulsed the UP output of the PD to the LOW state. The question now is: when will the signal $u_2(t)$ perform a negative transition? Because the PLL is not yet locked, the signal $u_2$ is not synchronized with the signal $u_1$. All we know is that the frequency of $u_2$ is $f_2 = \omega_2/2\pi$, or the period of $u_2$ is $T_2 = 1/f_2$. Consequently, we can say that the next negative transition of the signal $u_2$ will occur within the time interval $0 \le t \le T_2$, i.e., not earlier than at $t = 0$ and not later than at $t = T_2$.

The probability of the next negative transition of $u_2$ occurring within the time interval $0 \le t \le T_2$ is therefore

$$w(t) = \frac{1}{T_2} \qquad 0 \le t \le T_2 \tag{D-1a}$$

and of course

$$w(t) = 0 \qquad t > T_2 \tag{D-1b}$$

The probability density function $w(t)$ is shown in curve $c$ of Fig. D-1. In the example of Fig. D-1 the period $T_2$ has been chosen greater than $T_1 = 1/f_1$, but less than $2T_1$. We can therefore distinguish two cases:

1. The next negative transition of $u_2$ occurs within the time interval $0 \le t \le T_1$. (This is depicted in curve $d$ of Fig. D-1.)

*Appendix to Sec. 4-1-2.

295

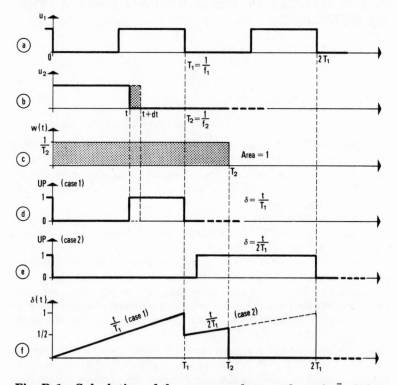

**Fig. D-1** Calculation of the average duty-cycle ratio $\bar{\delta}$ of the phase-detector output signal during the pull-in process (*a*) Waveform of the reference signal $u_1$. (*b*) Waveform of the VCO output signal $u_2$. (*c*) Plot of the probability $w(t)$ of the occurrence of the next negative-going transient of $u_2(t)$. (*d*) Waveform of the UP signal of the type 4 PD in case 1. [The negative edge of $u_2(t)$ occurs in the time interval $0 \leq t \leq T_1$.] (*e*) Same waveform as in (*d*), but for case 2 [The negative edge of $u_2(t)$ occurs in the time interval $T_1 \leq t \leq T_2$.] (*f*) Plot of the duty-cycle ratio $\delta(t)$ of the PD's UP signal against time.

2. The next negative transition of $u_2$ occurs within the time interval $T_1 \leq t \leq T_2$. (This is depicted in curve *e* of Fig. D-1.)

Remember now that the average output signal $\overline{u_d}$ of this PD is equal to the duty-cycle ratio of the $u_2$ waveform. The duty-cycle ratio, later to be called $\delta(t)$, is now calculated as a function of time when the next negative transition of $u_2$ occurs. In case 1 the duty-cycle ratio is evidently given by $\delta(t) = t/T_1$. If the next negative transition occurs immediately after $t = 0$, $\delta(t)$ is near zero, but if the next negative transition occurs shortly before $t = T_1$, $\delta(t)$ becomes almost unity. This is shown by curve $f$ in Fig. D-1.

What about case 2, however, where the next negative transition of $u_2$ occurs *after* $t = T_1$? If this happens shortly after $t = T_1$, $\delta$ is slightly more than 50 percent because the UP output will be set LOW only after $t = 2T_1$; $\delta$ approaches unity again if the next negative transition of $u_2$ occurs at a time $t$ slightly less than $2T_1$. In case 2 the duty-cycle ratio $\delta$ is therefore given by $\delta(t) = t/2T_1$.

We now are in a position to calculate the average duty-cycle ratio of the UP signal for the situation where $T_1 < T_2 < 2T_1$, i.e. for the case where $T_2$ is longer than $T_1$ but shorter than $2T_1$. Based on elementary statistics, we have

$$\bar{\delta} = \int_0^{T_2} w(t)\delta(t)\ dt \tag{D-2}$$

As seen from curve $f$ in Fig. D-1, the integration has to be performed in two steps,

$$\bar{\delta} = \int_0^{T_1} w(t)\,\frac{t}{T_1}\ dt + \int_{T_1}^{T_2} w(t)\,\frac{t}{2T_1}\ dt$$

Consider now that in the most general case $T_2$ is not restricted to the range $T_1 < T_2 < 2T_2$, but rather can have values between $T_1$ and $nT_1$, where $n$ is an arbitrarily large integer. The function $\delta(t)$ then no longer consists of just two sawtooth-like waveforms, but rather has $n$ segments, as shown in Fig. D-2. When integrating Eq. (D-2), we consequently obtain $n$ different terms,

$$\bar{\delta} = \int_0^{T_1} w(t)\,\frac{t}{T_1}\ dt + \int_{T_1}^{2T_1} w(t)\,\frac{t}{2T_1} + \cdots + \int_{(n-1)T_1}^{T_2} w(t)\,\frac{t}{nT_1}\ dt \tag{D-3}$$

If this expression is worked out, we obtain

$$\bar{\delta} = \frac{f_2}{2f_1}\left(n - \sum_{i=1}^n \frac{1}{i}\right) + \frac{f_1}{2nf_2} \qquad f_1 > f_2 \tag{D-4}$$

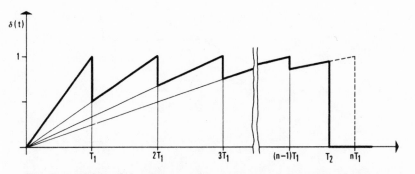

**Fig. D-2**  Plot of the duty-cycle ratio $\delta(t)$ against time for the most general case. [The negative edge of $u_2(t)$ occurs at any time in the time interval $0 \le t \le nT_1$, where $n$ is an arbitrary positive integer.]

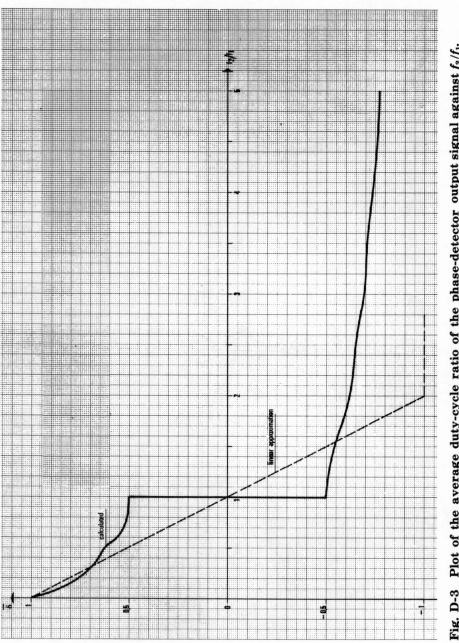

Fig. D-3 Plot of the average duty-cycle ratio of the phase-detector output signal against $f_2/f_1$.

In Eq. (D-4) $n$ is given by $n = \text{Int } (f_1/f_2) + 1$, that is, $n$ is the next integer which is greater than $f_1/f_2$ (for example, if $f_1/f_2 = 8.36$, $n = 9$). Equation (D-4), plotted graphically in Fig. D-3, was derived assuming $f_2 < f_1$. Hence this equation yields the curve only in the range $0 \le f_2/f_1 \le 1$. The remaining portion of the curve is obtained by some additional arguments. It is easily shown that for the case $f_2 > f_1$ the following identity holds true:

$$\bar{\delta}\left(\frac{f_2}{f_1}\right)\bigg|_{\text{upper tree}} = -\bar{\delta}\left(\frac{f_1}{f_2}\right)\bigg|_{\text{lower tree}} \tag{D-5}$$

If $f_2/f_1$ approaches unity from the left, $\bar{\delta}$ becomes $+0.5$; if $f_2/f_1$ approaches unity from the right, however, $\bar{\delta}$ becomes $-0.5$.

The rather complicated function $\bar{\delta}(f_2/f_1)$ as shown in Fig. D-3 does not prove suitable for the calculation of the pull-in process. A rough approximation becomes possible if the curve shown in Fig. D-3 is replaced by a straight line (see the dashed line in Fig. D-3). For this approximation one gets

$$\bar{\delta} = \frac{f_1 - f_2}{f_1} = \frac{\omega_1 - \omega_2}{\omega_1} \tag{D-6}$$

The average output signal of the PD $\overline{u_d}$ can now be calculated. If the supply voltage for the PD is $U_B$, its output signal is switched between the levels 0 and $U_B$. Consequently the average output signal $\overline{u_d}$ is

$$\overline{u_d} = \bar{\delta}U_B = \frac{U_B}{\omega_1}(\omega_1 - \omega_2) \tag{D-7}$$

Thus $\overline{u_d}$ is proportional to the frequency offset $\Delta\omega = \omega_1 - \omega_2$, and $U_B/\omega_1$ can be defined as the gain of the phase detector in the unlocked state, $K'_d$:

$$K'_d = \frac{U_B}{\omega_1} \tag{D-8a}$$

If the PLL is operating near its center frequency, $\omega_1 \approx \omega_0$, and Eq. (D-8a) becomes

$$K'_d \approx \frac{U_B}{\omega_0} \tag{D-8b}$$

[compare also with Eq. (4-15)].

The gain factor $K'_d$ should not be confused with the previously introduced phase-detector gain $K_d$ (in the locked state of the PLL), as defined in Eq. (2-3).

# *Appendix E*
## *THE PULL-OUT RANGE* $\omega_{PO}$ OF THE DIGITAL PLL[*]

As explained in Sec. 4-1-3, a DPLL containing a type 4 PD unlocks if the phase error $\theta_e$ exceeds $2\pi$. The pull-out range $\Delta\omega_{PO}$ of this type of DPLL is consequently the frequency step $\Delta\omega$ that causes the peak phase error to reach the value of $2\pi$.

As shown in Sec. 4-1-3, the transient response of this type of DPLL is calculated in the same way as the transient response of the linear PLL to a frequency step applied to its reference input. Consequently the phase error is given by Eqs. (C-5). When $\zeta < \omega$ we have

$$\theta_e(t) = \frac{\Delta\omega}{\omega_n} \left[ \frac{1}{\sqrt{1 - \zeta^2}} \sin(\sqrt{1 - \zeta^2}\,\omega_n t) \right] \exp(-\zeta\omega_n t) \tag{E-1a}$$

and when $\zeta > 1$ we obtain

$$\theta_e(t) = \frac{\Delta\omega}{\omega_n} \left[ \frac{1}{\sqrt{\zeta^2 - 1}} \sinh\sqrt{\zeta^2 - 1}\,\omega_n t \right] \exp(-\zeta\omega_n t) \tag{E-1b}$$

We will calculate the time $t_K$ at which $\theta_e$ has a maximum. This computation has to be performed separately for the two cases $\zeta < 1$ and $\zeta > 1$. For $\zeta < 1$, the derivation is as follows. Putting the first derivative of $\theta_e(t)$ zero yields

$$0 = \frac{d\theta_e}{dt} = \frac{\Delta\omega}{\omega_n} \frac{\exp(-\zeta\omega_n t_K)}{\sqrt{1 - \zeta^2}} \left[ (-\zeta\omega_n) \sin(\sqrt{1 - \zeta^2}\,\omega_n t_K) \right.$$
$$\left. + \sqrt{1 - \zeta^2}\,\omega_n \cos(\sqrt{1 - \zeta^2}\,\omega_n t_K) \right] \tag{E-2}$$

Solving for $t_K$ yields

$$\omega_n t_K = \frac{1}{\sqrt{1 - \zeta^2}} \tan^{-1} \frac{\sqrt{1 - \zeta^2}}{\zeta} \tag{E-3}$$

If this is substituted into Eq. (E-1a), we obtain

$$\theta_{e\,max} = \frac{\Delta\omega}{\omega_n} \exp\left[ -\frac{\zeta}{\sqrt{1 - \zeta^2}} \tan^{-1} \frac{\sqrt{1 - \zeta^2}}{\zeta} \right] \tag{E-4}$$

If $\theta_{e\,max}$ is now set to $2\pi$, $\Delta\omega$ becomes the pull-out range $\Delta\omega_{PO}$,

$$\frac{\Delta\omega_{PO}}{\omega_n} = 2\pi \exp\left[ \frac{\zeta}{\sqrt{1 - \zeta^2}} \tan^{-1} \frac{\sqrt{1 - \zeta^2}}{\zeta} \right] \quad \text{for} \quad \zeta < 1 \tag{E-5a}$$

[*]Appendix to Sec. 4-1-3.

For $\zeta > 1$ an analog computation yields

$$\frac{\Delta\omega_{PO}}{\omega_n} = 2\pi \exp\left[\frac{\zeta}{\sqrt{\zeta^2 - 1}} \tanh^{-1}\frac{\sqrt{\zeta^2 - 1}}{\zeta}\right] \quad \text{for} \quad \zeta > 1 \qquad \text{(E-5b)}$$

The pull-out range $\Delta\omega_{PO}$ is plotted in Fig. 4-6 as a function of the damping factor $\zeta$. If a type 3 PD is used instead of a type 4, the pull-out range is reduced by a factor of 2.

## Appendix F
## A PRIMER ON LAPLACE TRANSFORMS FOR ELECTRONIC ENGINEERS

### F-1 TRANSFORMS ARE THE ENGINEER'S TOOLS

Trying to solve electronic problems without using the Laplace transform is similar to traveling through a foreign country with a globe instead of a map (Fig. F-1). An engineer who tries to find the transient response of an electric network to an impulse function by solving differential equations, for example, certainly is working with inadequate tools, (see Fig. F-2). The engineer familiar with the techniques of the Laplace transform may find a solution very quickly, as shown in Fig. F-3.

A map images a three-dimensional object to a plane. Every spatial point of the three-dimensional object is represented by a unique point in the plane of the map. Things are similar, though different, in the case of the Laplace transform. Here a function in the *time domain* (such as an electric signal) is transformed to another function in the *complex frequency domain*. The trouble with the Laplace transform starts right here: even an electronic hobbyist can imagine what the frequency spectrum of an electric signal is, but what is *complex frequency?*

One need not be a cow to know what milk is, but it is surely easier to understand the term complex frequency if we first consider *real* frequency spectra. The Laplace transform is a more general form of the Fourier transform; in other words, the Fourier transform is a special case of the Laplace transform. In this context it may be easier to start with the special case before proceeding to the more general one.

The Fourier transform is explained best by first looking at a periodic signal such as a square wave (Fig. F-4). The angular frequency of this signal is assumed to be $\omega_0$. This square wave can now be thought to be composed of a(n) (infinite) number of sine-wave signals having frequencies $\omega_0$ (fundamental frequency), $2\omega_0$, $3\omega_0$, ... (harmonics). Mathematically, any periodic signal $f(t)$ having a repetition frequency of $\omega_0$ can be written as a sum of its harmonics,

$$f(t) = \sum_{n=-\infty}^{+\infty} F(jn\omega_0) \exp(jn\omega_0 t) \tag{F-1}$$

The so-called Fourier coefficients $F(jn\omega)$ are calculated from

$$F(jn\omega_0) = \frac{1}{T} \int_{-T/2}^{+T/2} f(t) \exp(jn\omega_0 t) \, dt \tag{F-2}$$

where $T$ is the period of the periodic signal $f(t)$, $T = 2\pi/\omega$, and $F(jn\omega_0)$ is the amplitude of the harmonic component with frequency $n\omega_0$. (The $j$ operator could

**Fig. F-1  Trying to solve electronic problems without using the Laplace transform is as cumbersome as traveling through a foreign country with a globe instead of a map.**

be omitted, but here we keep this term because any Fourier coefficient $F$ is always a function of $j\omega$ and never of $\omega$ alone.) Note that the Fourier coefficients $F(jn\omega_0)$ are generally complex numbers; We can therefore write

$$F(jn\omega_0) = |F(jn\omega_0)| \exp(j\phi_n) \tag{F-3}$$

where $|F(jn\omega_0)|$ is the amplitude and $\phi_n$ the phase of $F(jn\omega_0)$.

When plotting the Fourier transform of a signal $f(t)$, the amplitude $|F(jn\omega_0)|$ and the phase $\phi_n$ are normally plotted against $\omega$; these functions are

**Fig. F-2  Looking at the transient response of electric networks without using the Laplace transform can be tricky ...**

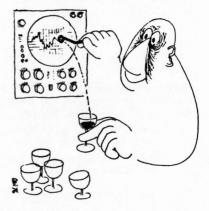

**Fig. F-3  ... but the engineer familiar with Laplace techniques may find the solution very quickly.**

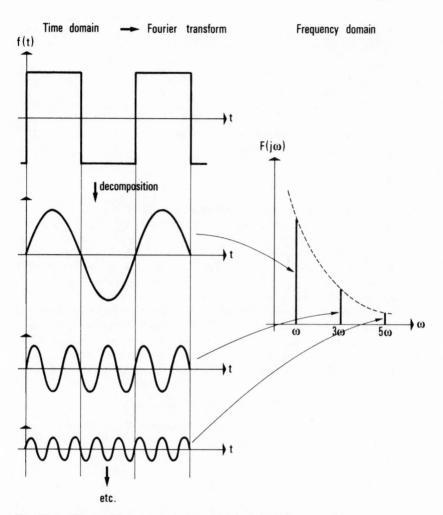

**Fig. F-4   The principle of the Fourier transform.**

called amplitude and phase spectra, respectively. In the case of periodic functions $f(t)$ the Fourier spectra become discrete, that is, the Fourier coefficients $F(jn\omega_0)$ are defined only at the discrete frequencies $\omega_0$, $2\omega_0$, $3\omega_0$, ....

Figure F-4 shows the amplitude spectrum $|F(jn\omega_0)|$ of a symmetrical square wave; for a symmetrical waveform it can be shown that the even harmonics disappear. Consequently, the Fourier spectrum of Fig. F-4 only shows lines at $\omega_0$, $3\omega_0$, $5\omega_0$, ....

In real life we find many signals which are not periodic, such as a single pulse. What about the Fourier transform of such signals? If a signal $f(t)$ is not periodic, we could state that its period approaches infinity, which means its fundamental frequency approaches zero. If the fundamental frequency approaches zero, the spectral lines in Eq. (F-1) come closer and closer, and finally the sum is replaced by an integral,

$$f(t) = \frac{1}{2\pi} \int_{-\infty}^{+\infty} F(j\omega) \, e^{j\omega t} \, dt \tag{F-4}$$

For an aperiodic signal $f(t)$ the Fourier spectrum $F(j\omega)$ becomes continuous; $F(j\omega)$ is called the *Fourier transform* of the signal $f(t)$. The Fourier transform is calculated in the same way as the Fourier coefficients $F(jn\omega)$ in Eq. (F-2). If $T$ approaches $\infty$ in this equation, we get

$$F(j\omega) = \int_{0}^{\infty} f(t) \, e^{-j\omega t} \, dt \tag{F-5}$$

This is the definition of the *continuous* Fourier transform.

Note that the Fourier transform $F(j\omega)$ for any given signal $f(t)$ can be calculated by evaluating the integral in Eq. (F-5). If the Fourier transform of a signal $f(t)$ is given first, the corresponding signal $f(t)$ in the time domain can be obtained by applying Eq. (F-4). This equation is also called the *inverse* Fourier transform. The Fourier transforms $F(j\omega)$ for the most important signal waveforms are tabulated in many reference books; see for example, Ref. 27.

The usefulness of the Fourier transform is limited by a serious drawback: if we try to evaluate the Fourier integral in Eq. (F-5), we find that the integral does *not converge* for most signals $f(t)$. You will find it impossible to find a solution of the Fourier integral by conventional methods, even for such a simple signal as $f(t) = \sin \omega t$. The Laplace transform offers a way to circumvent this problem.

## F-2 A LAPLACE TRANSFORM IS THE KEY TO SUCCESS

Imagine we would like to know the Fourier transform of an extremely simple signal, $f(t) = \sin \omega_0 t$. Using the definition of the Fourier transform, Eq. (F-5), we get

$$F(j\omega) = \int_{0}^{\infty} f(t) \, e^{-j\omega t} \, dt = \int_{0}^{\infty} \sin \omega_0 t \, e^{-j\omega t} \, dt$$

Using an integral table we find for $F(j\omega)$

$$F(j\omega) = \frac{-j\omega \, e^{-j\omega t} \sin \omega_0 t - \omega_0 \, e^{-j\omega t} \cos \omega_0 t}{\omega_0^2 - \omega^2} \bigg|_{0}^{\infty}$$

Introducing the limits of integration (here 0 and $\infty$) yields

$$F(j\omega) = \frac{-j\omega \, e^{-j\infty} \sin \infty - \omega_0 \, e^{-j\infty} \cos \infty + \omega_0}{\omega_0^2 - \omega^2}$$

But what is $\sin \infty$, and what is $\cos \infty$? Both functions are periodic and are within a range of $-1$ to $+1$; hence $\sin \infty$ and $\cos \infty$ are not defined.

The evaluation of the Fourier integral would be simpler if the function $f(t)$ would approach zero for very large values of $t$. The solution of the Fourier integral becomes possible if $f(t)$ is multiplied by a damping function $e^{-\sigma t}$, where $\sigma$ is a positive real number:

$$F'(j\omega, \sigma) = \int_0^\infty [f(t) \, e^{-\sigma t}] \, e^{-j\omega t} \, dt \qquad \text{(F-6)}$$

The notation $F'(j\omega, \sigma)$ is chosen to emphasize that now $F'$ is also dependent on $\sigma$. The prime is furthermore chosen to differentiate $F'$ from the Fourier integral in Eq. (F-5). The product $[f(t) \, e^{-\sigma t}]$ approaches zero for large $t$. This is true even when $f(t)$ is an exponential function $f(t) = e^{at}$ with positive $a$. If $\sigma$ is chosen larger than $a$, the product approaches zero for $t \to \infty$. The modified Fourier integral in Eq. (F-6) is now easily evaluated. Let us perform the calculation for the previous example $f(t) = \sin \omega_0 t$. We then have

$$F'(j\omega, \sigma) = \int_0^\infty \sin \omega_0 t \, e^{-\sigma t} \, e^{-j\omega t} \, dt$$

Again using an integral table, we get

$$F'(j\omega, \sigma) = \frac{-(\sigma + j\omega) \, e^{-(\sigma+j\omega)t} \sin \omega_0 t - \omega_0 \, e^{-(\sigma+j\omega)t} \cos \omega_0 t}{\omega_0^2 + (\sigma + j\omega)^2} \Bigg|_0^\infty$$

If the limits of integration are now inserted, the terms

$$e^{-(\sigma+j\omega)\infty}$$

become zero for positive $\sigma$. Then $F'(j\omega)$ becomes

$$F'(j\omega, \sigma) = \frac{\omega_0}{\omega_0^2 + (\sigma + j\omega)^2}$$

Of course $F'(j\omega)$ contains the damping factor $\sigma$. The Fourier integral $F(j\omega)$ is now obtained simply by letting $\sigma \to 0$:

$$F(j\omega) = \lim_{\sigma \to 0} F'(j\omega, \sigma) = \frac{\omega_0}{\omega_0^2 - \omega^2}$$

If $\sigma$ is set at zero in Eq. (F-6), this equation is transformed into Eq. (F-5). *Thus the Fourier transform of any function $f(t)$ is obtained by first introducing a damping function $e^{-\sigma t}$ evaluating the integral [Eq. (F-6)] for $\sigma > 0$ and finally letting $\sigma \to 0$.*

Let us now see what the Laplace transform really is. Equation (F-6) can also be written in a different form;

$$F'(j\omega, \sigma) = \int_0^\infty f(t)[e^{-\sigma t} \, e^{-j\omega t}] \, dt$$

In contrast to Eq. (F-6), we now combine the damping function $e^{-\sigma t}$ with the exponential function $e^{-j\omega t}$. This can be written as

$$F'(j\omega, \sigma) = \int_0^\infty f(t) \, e^{-(\sigma+j\omega)t} \, dt$$

We now define

$$s = \sigma + j\omega$$

and call the new variable $s$ complex frequency. Because the two variables $\sigma$ and $\omega$ appear only in the form $\sigma + j\omega$, $F'(j\omega, \sigma)$ can be written as $F(\sigma + j\omega) = F(s)$. We then have

$$F(s) = \int_0^\infty f(t) \, e^{-st} \, dt \qquad\qquad\qquad \text{(F-7)}$$

This is the definition of the Laplace transform.

In the case of the Fourier transform, $F(j\omega)$ was a *complex* function of the *real* variable $\omega$. To plot $F(j\omega)$ against $\omega$ we had to plot amplitude $|F(j\omega)|$ and phase $\phi(j\omega)$ as a function of $\omega$. (This plot is comonly called a Bode diagram.) In the case of the Laplace transform, however, $F(s)$ is a *complex* function of the *complex* variable $s$. Plotting $F(s)$ is much more difficult than plotting $F(j\omega)$. To plot $F(s)$, a relief of the amplitude $|F(s)|$ and of the phase $\phi(s)$ could be constructed; a relief of $|F(s)|$ is shown in Fig. F-5.

The construction of such a relief is a cumbersome procedure. As we shall see later, it is sufficient to know the locations of some singular points of $F(s)$ only, namely the *poles* and *zeros* of $F(s)$. A pole is a point in the $s$ plane where $F(s)$ becomes infinity, and a zero is a point in the $s$ plane where $F(s)$ is zero. The so-called pole-zero plot (Fig. F-6) shows the locations in the $s$ plane where $F(s)$ has poles (•) or zeros (○). (Note that not every function $F(s)$ necessarily has poles or zeros.)

We see immediately that the Fourier transform [Eq. (F-5)] is a special case

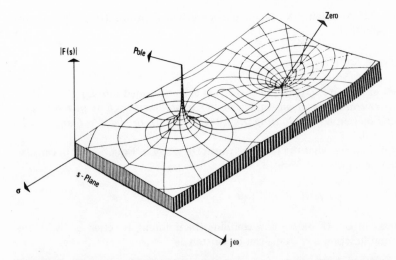

**Fig. F-5** **The magnitude of the Laplace transform $F(s)$ plotted as a function of complex frequency $s$ yields a relief. This illustration shows a pole and a zero of the transfer function $F(s)$.**

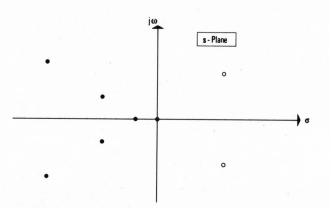

**Fig. F-6 Plot of the poles and zeroes of a given transfer function $F(s)$.**

of the Laplace transform [Eq. (F-7)]. The Fourier transform $F(j\omega)$ is simply obtained by finding the Laplace transform $F(s)$ and then putting $\sigma = 0$

$$F(j\omega) = \lim_{\sigma \to 0} F(s), \qquad s = \sigma + j\omega \tag{F-8}$$

In other words, $F(j\omega)$ is equal to the function $F(s)$ on the imaginary axis of the $s$ plane.

Equation (F-7) shows us how to calculate the Laplace transform from a given function $f(t)$ in the time domain. As was the case with the Fourier transform, it is also possible to calculate $f(t)$ if $F(s)$ is given first. This transformation is called *inverse* Laplace transform and is defined by

$$f(t) = \frac{1}{2\pi j} \int_{c-j\infty}^{c+j\infty} F(s)\, e^{-st}\, ds \tag{F-9}$$

where $c$ is a real constant. As is seen from Eq. (F-9) the path of integration is a line parallel to the imaginary axis.[3,4,20]

In most cases, computation of the Laplace transform or the inverse Laplace transform by evaluating Eqs. (F-7) and (F-9) is no longer necessary since transformation tables are available (see Table F-1). To conclude this section, let us introduce some convenient abbreviations:

$$F(s) = \mathcal{L}\{f(t)\} \tag{F-10a}$$

$$f(t) = \mathcal{L}^{-1}\{F(s)\} \tag{F-10b}$$

Equation (F-10a) says that $F(s)$ is the Laplace transform of the signal $f(t)$; Eq. (F-10b) states that $f(t)$ is the inverse Laplace transform of $F(s)$.

### F-3 A NUMERICAL EXAMPLE OF THE LAPLACE TRANSFORM

To see how the Laplace transform $F(s)$ of a signal $f(t)$ could be obtained by evaluating Eq. (F-7), let us calculate an example.

**Table F-1. Laplace transform $f(t) = 0$ for $t < 0$**

| | Signal $f(t)$ | Laplace transform $F(s)$ |
|---|---|---|
| **GENERAL RULES** | | |
| Differentiation | $\dfrac{df(t)}{dt}$ | $sF(s) - f(0)$ |
| Integration | $\displaystyle\int_0^t f(t)\,dt$ | $\dfrac{F(s)}{s} + \dfrac{f^{(-1)}(0)}{s}$ |
| Time delay | $f(t - \tau)$ | $F(s)e^{-s\tau}$ |
| Multiplication by constant | $kf(t)$ | $kF(s)$ |
| Convolution | $f_1(t) \cdot f_2(t)$ | $F_1(s) * F_2(s)$ |
| | $f_1(t) * f_2(t)$ | $F_1(s) \cdot F_2(s)$ |
| **FUNCTIONS** | $0$ | $0$ |
| Unit step | $u(t)$ | $\dfrac{1}{s}$ |
| Delta function | $\delta(t)$ | $1$ |
| Ramp function | $t$ | $\dfrac{1}{s^2}$ |
| Parabola | $\dfrac{t^2}{2}$ | $\dfrac{1}{s^3}$ |
| Polynomial in $t$ | $\dfrac{t^{n-1}}{(n-1)!}$ | $\dfrac{1}{s^n}$, $n > 0$ (can be fractional) |
| Exponential functions | $e^{\alpha t}$ | $\dfrac{1}{s - \alpha}$ |
| | $\dfrac{1}{\alpha}(e^{\alpha t} - 1)$ | $\dfrac{1}{s(s - \alpha)}$ |
| | $\dfrac{1}{\alpha}(1 - e^{-\alpha t})$ | $\dfrac{1}{s(s + \alpha)}$ |
| | $\dfrac{t^n - 1}{(n - 1)!e^{\alpha t}}$ | $\dfrac{1}{(s - \alpha)^n}$, $n > 0$ (can be fractional) |

|  | Signal $f(t)$ | Laplace transform $F(s)$ |
|---|---|---|
| Trigonometric functions | $\dfrac{1}{\alpha}\sin \alpha t$ | $\dfrac{1}{s^2 + \alpha^2}$ |
|  | $\cos \alpha t$ | $\dfrac{s}{s^2 + \alpha^2}$ |
|  | $\dfrac{1}{\alpha^2}(1 - \cos \alpha t)$ | $\dfrac{1}{s(s^2 + \alpha^2)}$ |
| Hyperbolic functions | $\dfrac{1}{\alpha}\sinh \alpha t$ | $\dfrac{1}{s^2 - \alpha^2}$ |
|  | $\cosh \alpha t$ | $\dfrac{s}{s^2 - \alpha^2}$ |
|  | $\dfrac{1}{\alpha^2}(\cosh \alpha t - 1)$ | $\dfrac{1}{s(s^2 - \alpha^2)}$ |
| Second-order systems (aperiodic) | $\dfrac{e^{\beta t} - e^{\alpha t}}{\beta - \alpha}$ | $\dfrac{1}{(s - \alpha)(s - \beta)}$ |
|  | $\dfrac{\beta e^{\beta t} - \alpha e^{\alpha t}}{\beta - \alpha}$ | $\dfrac{s}{(s - \alpha)(s - \beta)}$ |
|  | $\dfrac{\beta e^{\alpha t} - \alpha e^{\beta t}}{\alpha\beta(\alpha - \beta)} + \dfrac{1}{\alpha\beta}$ | $\dfrac{1}{s(s - \alpha)(s - \beta)}$ |
| Second-order systems (periodic) | $\dfrac{e^{-\zeta\omega_n t}\sin \sqrt{1 - \zeta^2}\,\omega_n t}{\sqrt{1 - \zeta^2}\,\omega_n}$ | $\dfrac{1}{s^2 + 2s\zeta\omega_n + \omega_n^2}$ |
|  | $\left[\cos \sqrt{1 - \zeta^2}\,\omega_n t - \right.$ $\left. \dfrac{\zeta}{\sqrt{1 - \zeta^2}}\sin \sqrt{1 - \zeta^2}\,\omega_n t\right] e^{-\zeta\omega_n t}$ | $\dfrac{s}{s^2 + 2s\zeta\omega_n + \omega_n^2}$ |
|  | $\omega_n e^{-\zeta\omega_n t}\left[\dfrac{2\zeta_2 - 1}{\sqrt{1 - \zeta^2}}\sin \sqrt{1 - \zeta^2}\,\omega_n t\right.$ $\left. - 2\zeta \cos \sqrt{1 - \zeta^2}\,\omega_n t\right]$ | $\dfrac{s^2}{s^2 + 2s\zeta\omega_n + \omega_n^2}$ |
|  | $\dfrac{1}{\omega_n^2}\left[1 - \left(\cos \sqrt{1 - \zeta^2}\,\omega_n t\right.\right.$ $\left.\left. + \dfrac{\zeta}{\sqrt{1 - \zeta^2}}\sin \sqrt{1 - \zeta^2}\,\omega_n t\right) e^{-\zeta\omega_n t}\right]$ | $\dfrac{1}{s(s^2 + 2s\zeta\omega_n + \omega_n^2)}$ |

**Table F-1. (continued)**

| | Signal $f(t)$ | Laplace transform $F(s)$ |
|---|---|---|
| Third-order systems | $\dfrac{e^{\alpha t} - [1 + (\alpha - \beta)t]e^{\beta t}}{(\alpha - \beta)^2}$ | $\dfrac{1}{(s - \alpha)(s - \beta)^2}$ |
| | $\dfrac{\alpha e^{\alpha t} - [\alpha + \beta(\alpha - \beta)t]e^{\beta t}}{(\alpha - \beta)^2}$ | $\dfrac{s}{(s - \alpha)(s - \beta)^2}$ |
| | $\dfrac{\alpha^2 e^{\alpha t} - [2\alpha - \beta + \beta(\alpha - \beta)t]\beta e^{\beta t}}{(\alpha - \beta)^2}$ | $\dfrac{s^2}{(s - \alpha)(s - \beta)^2}$ |
| | $\dfrac{(\beta - \gamma)e^{\alpha t} + (\gamma - \alpha)e^{\beta t} + (\alpha - \beta)e^{\gamma t}}{(\alpha - \beta)(\beta - \gamma)(\gamma - \alpha)}$ | $\dfrac{1}{(s - \alpha)(s - \beta)(s - \gamma)}$ |
| Various functions | $\dfrac{\alpha \sin \beta t - \beta \sin \alpha t}{\alpha\beta(\alpha^2 - \beta^2)}$ | $\dfrac{1}{(s^2 + \alpha^2)(s^2 + \beta^2)}$ |
| | $\dfrac{\cos \beta t - \cos \alpha t}{\alpha^2 - \beta^2}$ | $\dfrac{s}{(s^2 + \alpha^2)(s^2 + \beta^2)}$ |
| | $\dfrac{1}{\sqrt{\pi t}}$ | $\dfrac{1}{\sqrt{s}}$ |
| | $2 \sqrt{\dfrac{t}{\pi}}$ | $\dfrac{1}{s\sqrt{s}}$ |
| | $\dfrac{n!}{(2n)!} \cdot \dfrac{4^n}{\sqrt{\pi}} t^{n-1/2}$ | $\dfrac{1}{s^n\sqrt{s}}$ |
| | $\dfrac{1}{\sqrt{\pi t}} e^{\alpha t}$ | $\dfrac{1}{\sqrt{s - \alpha}}$ |
| Error functions (erf) | $\dfrac{2}{\sqrt{\alpha\pi}} \displaystyle\int_0^{\sqrt{\alpha t}} e^{-\zeta^2}\, d\zeta$ | $\dfrac{1}{s\sqrt{s + \alpha}}$ |
| | $\dfrac{2e^{-\alpha t}}{\sqrt{\pi(\beta - \alpha)}} \displaystyle\int_0^{\sqrt{(\beta-\alpha)t}} e^{-\zeta^2}\, d\zeta$ | $\dfrac{1}{(s + \alpha)\sqrt{s + \beta}}$ |
| | $\dfrac{e^{-\alpha t}}{\sqrt{\pi t}} + 2 \sqrt{\dfrac{\alpha}{\pi}} \displaystyle\int_0^{\sqrt{\alpha t}} e^{-\zeta^2}\, d\zeta$ | $\dfrac{\sqrt{s + \alpha}}{s}$ |
| Bessel function of zero order | $I_0(\alpha t)$ | $\dfrac{1}{\sqrt{s^2 + \alpha^2}}$ |
| Modified Bessel function of zero order | $J_0(\alpha t)$ | $\dfrac{1}{\sqrt{s^2 - \alpha^2}}$ |

Assume $f(t)$ is the unit step function

$$f(t) = u(t)$$

as plotted in Fig. F-7. For the Laplace transform $F(s)$ we have, according to Eq. (F-7),

$$F(s) = \int_0^\infty u(t)\, e^{-st}\, dt = \int_0^\infty e^{-st}\, dt = \frac{e^{-st}}{-s}\bigg|_0^\infty$$

$$= -\frac{1}{s}(e^{-\infty} - e^0) = -\frac{1}{s}(0 - 1) = \frac{1}{s}$$

that is,

$$F(s) = \frac{1}{s}$$

$F(s)$ has one single pole at $s = 0$, that is, at the origin of the $s$ plane. Furthermore, the Fourier transform of the unit step is obtained immediately by putting $s = j\omega$;

$$F(j\omega) = \frac{1}{j\omega}$$

Evaluating the Laplace integral in Eq. (F-7) for the most common test signals $f(t)$, yields Table F-1.

### F-4 SOME BASIC PROPERTIES OF THE LAPLACE TRANSFORM

For practical applications it is useful to be familiar with some basic properties of the Laplace transform. The most important ones are discussed below.

### F-4-1 Addition Theorem

The Laplace transform as defined by Eq. (F-7) is linear with respect to $f(t)$. If two signals $f_1(t)$ and $f_2(t)$ are given, the Laplace transform of the sum of $f_1 + f_2$ is therefore equal to the sum of the individual Laplace transforms $F_1(s)$ and $F_2(s)$;

$$\mathcal{L}\{f_1(t) + f_2(t)\} = \mathcal{L}\{f_1(t)\} + \mathcal{L}\{f_2(t)\} = F_1(s) + F_2(s) \tag{F-11}$$

### F-4-2 Multiplication by a Constant Factor $k$

For the same reason we have

$$\mathcal{L}\{k(f(t))\} = k\mathcal{L}\{f(t)\} = kF(s) \tag{F-12}$$

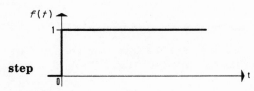

**Fig. F-7 Plot of the unit step function against time.**

If the signal $f(t)$ is multiplied by a constant factor $k$, the Laplace transform $F(s)$ is simply multiplied by the same factor.

### F-4-3 Multiplication of Signals

If two signals $f_1(t)$ and $f_2(t)$ are multiplied together in the time domain, the Laplace transform of $f_1(t) \cdot f_2(t)$ is *not* given by multiplying the individual Laplace transforms $F_1(s) = \mathcal{L}\{f_1(t)\}$ and $F_2(s) = \mathcal{L}\{f_2(t)\}$. Thus

$$\mathcal{L}\{f_1(t) \cdot f_2(t)\} \neq F_1(s) \cdot F_2(s)$$

The operation in the complex frequency domain which corresponds to the multiplication in the time domain is much more complicated; it is called *complex convolution*.[20] We will not consider this operation in detail here, but will define it for completeness:

$$\mathcal{L}\{f_1(t) \cdot f_2(t)\} = \frac{1}{2\pi j} \oint F_1(\chi) \cdot F_2(s - \chi) \, dx \qquad \text{(F-13)}$$

where

$$F_1(s) = \mathcal{L}\{f_1(t)\}$$
$$F_2(s) = \mathcal{L}\{f_2(t)\}$$

and $\chi$ is an auxiliary complex variable. The integral on the right-hand side of Eq. (F-13) is called a *complex convolution integral*. The contour of integration is a closed loop and must be chosen on the basis of mathematical considerations[20] that will not be discussed here. For the complex convolution integral the simplified form

$$\frac{1}{2\pi j} \oint F_1(\chi) \cdot F_2(s - \chi) \, dx = F_1(s) * F_2(s)$$

is often used.

Now we determine which operation in the time domain corresponds to a multiplication in the complex frequency domain. The result is given without a mathematical derivation:

$$\mathcal{L}^{-1}\{F_1(s) \cdot F_2(s)\} = \int_0^t f_1(\tau) f_2(t - \tau) \, dt \qquad \text{(F-14)}$$

The integral on the right-hand side of Eq. (F-14) is called *convolution* and is often written as

$$\int_0^t f_1(\tau) f_2(t - \tau) \, dt = f_1(t) * f_2(t)$$

### F-4-4 Delay in the Time Domain

A signal $f(t)$ and its Laplace transform $F(s)$ are given. The signal is now delayed by the time interval $\tau$ (Fig. F-8). What is the Laplace transform of the delayed

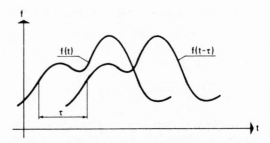

**Fig. F-8  Plot of a function $f(t)$ displaced by a time delay $\tau$.**

signal $f(t - \tau)$? The derivation[4] yields

$$\mathcal{L}\{f(t - \tau)\} = F(s)\, e^{-s\tau} \tag{F-15}$$

We shall calculate the Laplace transform of a rectangular pulse (Fig. F-9) as a numerical example. The pulse is first decomposed into two unit step functions, one starting at $t = 0$ and the other being delayed by the time interval $\tau$. The Laplace transform of the first unit step function is given by $1/s$, that of the second by $-(1/s) \cdot e^{-s\tau}$. Hence the Laplace transform of the pulse becomes

$$F(s) = \frac{1 - e^{-s\tau}}{s}$$

It is interesting to see that the Laplace transform describes a function with one single expression, whereas three separate equations would be required to describe the pulse in the time domain,

$$f(t) = \begin{cases} 0 & t < 0 \\ 1 & 0 \le t \le \tau \\ 0 & t > \tau \end{cases}$$

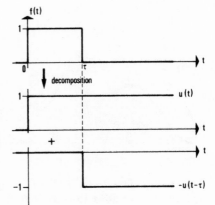

**Fig. F-9 If the Laplace transform of a single square-wave pulse (top trace) has to be calculated, the pulse is first decomposed into two unit step functions (center and bottom traces).**

315

### F-4-5 Differentiation and Integration in the Time Domain

A signal $f(t)$ and its Laplace transform $F(s)$ are given. In many cases we are interested to know the Laplace transform of the derivative $df/dt$ or of the integral $\int_0^t f(t)\,dt$. The Laplace transform of the derivative is[4]

$$\mathcal{L}\left\{\frac{df(t)}{dt}\right\} = sF(s) - f(0) \tag{F-16}$$

where $f(0)$ is the value of the signal $f(t)$ at $t = 0$. A differentiation in the time domain corresponds to a multiplication by $s$ in the complex frequency domain. The second term in Eq. (F-16) can be dropped if $f(0)$ is zero.

A similar rule applies to integration. The Laplace transform of the integral of a function $f(t)$ is[4]

$$\mathcal{L}\left\{\int_0^t f(t)\,dt\right\} = \frac{F(s)}{s} - \frac{f^{(-1)}(0)}{s} \tag{F-17}$$

The term $f^{(-1)}(0)$ is by definition the integral of $f(t)$ over the time interval between $-\infty$ and 0:

$$f^{(-1)}(0) = \int_{-\infty}^{0} f(t)\,dt$$

$f^{(-1)}$ is equal to the shaded area in Fig. F-10. If $f(t)$ is zero for $t < 0$, the second term in Eq. (F-17) can be dropped. Integration in the time domain corresponds to a division by $s$ in the complex frequency domain.

Let us now apply the rule of integration to an example (Fig. F-11). This figure shows a unit step function $u(t)$ and its first and second integrals. The first integral is a ramp function given by $f(t) = t$ in the time domain, and the second integral is a parabola given by $f(t) = t^2/2$. Because $f(t) = 0$ for $t < 0$ in the case of the unit step function, $f^{(-1)}$ is zero. The Laplace transform of the unit step has been shown to be $F(s) = 1/s$ (see Sec. F-3 or Table F-1). The Laplace transform of the ramp function (second row in Fig. F-11) is therefore obtained by a division by $s$,

$$F(s) = \frac{1}{s^2}$$

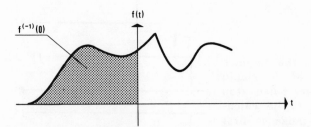

**Fig. F-10  The term $f^{(-1)}$ (0) is given by the shaded area under the curve $f(t)$.**

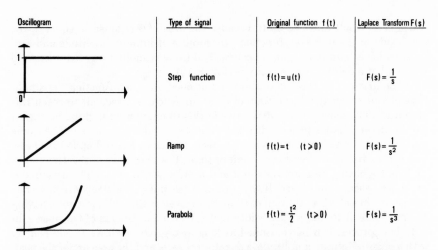

| Oscillogram | Type of signal | Original function $f(t)$ | Laplace Transform $F(s)$ |
|---|---|---|---|
| | Step function | $f(t) = u(t)$ | $F(s) = \dfrac{1}{s}$ |
| | Ramp | $f(t) = t \quad (t > 0)$ | $F(s) = \dfrac{1}{s^2}$ |
| | Parabola | $f(t) = \dfrac{t^2}{2} \quad (t > 0)$ | $F(s) = \dfrac{1}{s^3}$ |

**Fig. F-11 Example of applying the rule of integration. The Laplace transforms of a unit step, a ramp, and a parabola are determined.**

Finally the Laplace transform of the parabola (third row in Fig. F-11) is given by

$$F(s) = \frac{1}{s^3}$$

Let us now work out an examples of the differentiation rule [refer to Eq. (F-16)]. The unit step function (see, for examples, Fig. F-7) has been used throughout this text. Its Laplace transform has been shown to be $F(s) = 1/s$. But what is the first derivative of the step function, and what is its Laplace transform?

The first derivative of the unit step is called the *delta function* $\delta(t)$ (or impulse function). For $t < 0$ the unit step function is 0. Hence its derivative is also 0 in this range. For $t > 0$ the unit step function is 1. Its derivative must therefore also be 0. At $t = 0$ the unit step function shows a transient from 0 to 1. Consequently its derivative, the delta function, must be infinite at $t = 0$. The delta function $\delta(t)$ is shown in Fig. F-12.

The amplitude of the delta function is $\infty$, the pulse width is 0. The area under the delta function is obtained by multiplying the amplitude by the pulse

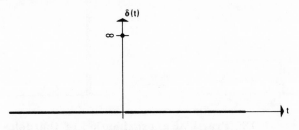

**Fig. F-12 Plot of the delta function $\delta(t)$.**

width. The result must be unity because the area of $\delta(t)$ must be equal to the amplitude of the unit step function. An impulse of infinite amplitude and zero pulse width is hard to imagine, of course. But there is another way to understand better what a delta function really is (Fig. F-13).

The first row of this figure shows a flattened unit step function. Its amplitude is still 1, but it takes a time of one unit (such as 1 second) to reach this amplitude. Consequently the derivative of this function is an impulse having an amplitude of 1 and a pulse width of 1. The area under the pulse is 1. Assume now that our unit step becomes a little steeper (second row in Fig. F-13); therefore it rises from 0 to 1 within 0.5 unit of time. The derivative of this function is an impulse having an amplitude of 2 and a pulse width of 0.5. The area under the pulse is still 1, of course. If the unit step becomes even steeper, it may rise from 0 to 1 in as little as 0.1 unit of time. Its derivative then will be a pulse having an amplitude of 10 and a pulse width of 0.1. Of course the area of the pulse still is 1. This process can be continued as long as desired, until we finally end up with a pulse of infinite amplitude, a duration of zero, and the area under the peak is still 1.

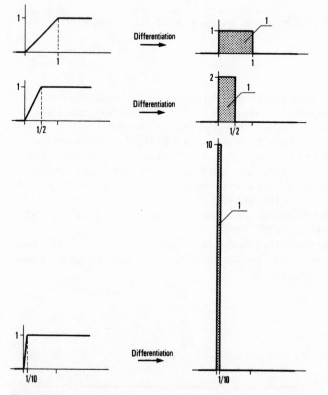

**Fig. F-13  Practical approximation of the delta function.**

What is the Laplace transform of the delta function now? The answer is very simple, because we only have to multiply the Laplace transform of the unit step function by $s$. The Laplace transform of the unit step has been shown to be

$$F(s) = \frac{1}{s} \quad \text{unit step}$$

Hence the Laplace transform of the delta function is

$$F(s) = 1$$

### F-4-6 The Initial- and Final-Value Theorems

In many cases where the Laplace transform $F(s)$ of a signal $f(t)$ is given, we are interested in knowing only the *initial value* $f(0)$ or the *final value* $f(\infty)$ of $f(t)$. The initial and final values $f(0)$ and $f(\infty)$, respectively, can be obtained immediately from the Laplace transform $F(s)$ without performing the inverse Laplace transform. The initial- and final-value theorems are given here without proof.[4] The initial-value theorem reads:

$$f(0) = \lim_{s \to \infty} sF(s) \tag{F-18}$$

The final-value theorem reads:

$$f(\infty) = \lim_{s \to 0} sF(s) \tag{F-19}$$

A numerical example is given to illustrate these theorems. The Laplace transform of a (hitherto) unknown signal $f(t)$ is given by

$$F(s) = \frac{1}{s(1 + sT)}$$

where $T$ is a time constant. What are the values of $f(0)$ and $f(\infty)$? Using the initial-value theorem [Eq. (F-18)], we get

$$f(0) = \lim_{s \to \infty} sF(s) = \lim_{s \to \infty} \frac{\cancel{s}}{\cancel{s}(1 + sT)} = 0$$

On the other hand, the final value, according to Eq. (F-19), is

$$f(\infty) = \lim_{s \to 0} sF(s) = \lim_{s \to 0} \frac{\cancel{s}}{\cancel{s}(1 + sT)} = 1$$

### F-5 USING THE TABLE OF LAPLACE TRANSFORMS

Table F-1 lists the Laplace transforms of the most commonly used signals $f(t)$. The table also includes some of the most important theorems of the Laplace transform. When using the table we should be aware that the Laplace transform $F(s)$ was obtained by integrating the Laplace integral of Eq. (F-7) over the time interval $0 \le t < \infty$. The values of the signal $f(t)$ at negative $t$ therefore did not contribute to $F(s)$. It is equivalent to state that $f(t)$ is effectively 0 for negative $t$.

This fact has an effect on the inverse Laplace transform. Performing the inverse Laplace transform for a given function $F(s)$ yields signal values $f(t)$ for positive $t$ only. If the Laplace transform of a signal $f(t)$ is given by, say,

$$F(s) = \frac{1}{s + a}$$

the table gives the corresponding signal $f(t)$ as

$$f(t) = e^{-at}$$

This holds true for positive $t$ only. For negative $t$, $f(t)$ has to be set so that $f(t) = 0$.

### F-6 APPLYING THE LAPLACE TRANSFORM TO ELECTRIC NETWORKS

The Laplace transform is the most effective tool for analyzing the transient response of electric networks. All linear electric devices, from electric motors to operational amplifiers, are modeled by a configuration of passive elements (resistor $R$, inductor $L$, and capacitor $C$) and active elements (voltage and current sources, controlled voltage and current sources). The transient response of such electric networks is analyzed in the time domain by writing the differential equations for the branch currents and voltages. Voltages and currents in the $R$, $L$, and $C$ elements are related by Ohm's law as follows: For resistors,

$$u(t) = Ri(t) \tag{F-20a}$$

for inductors,

$$u(t) = L\frac{di}{dt} \tag{F-20b}$$

and for capacitors,

$$u(t) = \frac{1}{C} \int i\, dt \tag{F-20c}$$

When analyzing a network in the complex frequency domain [using the rules of differentiation, Eq. (F-16), and of integration, Eq. (F-17)], we obtain, for resistors,

$$U(s) = RI(s) \tag{F-21a}$$

for inductors,

$$U(s) = L[sI(s) + i(0)] \tag{F-21b}$$

and for capacitors,

$$U(s) = \frac{1}{C}\left[\frac{I(s)}{s} + \frac{i^{(-1)}(0)}{s}\right] \tag{F-21c}$$

If the initial current $i(0)$ in an inductor $L$ is zero, the second term in Eq. (F-21$b$) is also zero. In this case we have

$$U(s) = sLI(s)$$

and the quotient

$$\frac{U(s)}{I(s)} = sL \qquad\qquad\qquad (\text{F-22}a)$$

can be defined as the *impedance* of the inductor. If the Laplace transform is replaced by the Fourier transform, $s$ is replaced by $j\omega$, and this expression becomes

$$\frac{U(j\omega)}{I(j\omega)} = j\omega L$$

which is the familiar ac impedance of the inductor known from the theory of alternating currents.

If the initial charge $i^{(-1)}/C$ in a capacitor $C$ is zero, the second term in Eq. (F-21$c$) is also zero. In this case we have

$$U(s) = \frac{1}{sC} I(s)$$

and the quotient

$$\frac{U(s)}{I(s)} = \frac{1}{sC} \qquad\qquad\qquad (\text{F-22}b)$$

is defined as the *impedance* of the capacitor. If the Laplace transform is again replaced by the Fourier transform, $s$ is replaced by $j\omega$, and Eq. (F-22$b$) becomes

$$\frac{U(j\omega)}{I(j\omega)} = \frac{1}{j\omega C}$$

which is the familiar ac impedance of the capacitor known from the theory of alternating currents.

Let us now apply the Laplace transform to the analysis of a simple electric network.

### Numerical Example: Transient Response of the Passive RC "Differentiator"

We want to find the transient response $u_2(t)$ of the differentiator in Fig. F-14$a$ on a single square-wave pulse $u_1(t)$. First we introduce the variables in the time domain on the right-hand side of Fig. F-14$a$. These variables are transformed into the complex frequency domain in Fig. F-14$b$. It is assumed that there is no initial charge on the capacitor.

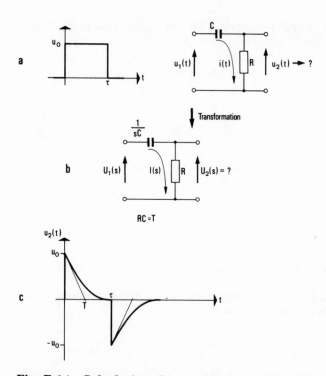

**Fig. F-14  Calculating the transient response of the passive *RC* differentiator using the Laplace transform. (*a*) Defining the variables in the time domain. (*b*) Introducing variables in the complex-frequency domain. (*c*) Plot of the output signal $u_2$ against time.**

*Solution*

We can now write the node and mesh equations of the network directly in the complex frequency domain. Making use of Eqs. (F-21*a*) and (F-22*b*), we have

$$U_1(s) = RI(s) + \frac{1}{sC} I(s)$$

$$U_2(s) = RI(s)$$

After elimination of $I(s)$ we obtain

$$U_2(s) = \frac{sRC}{1 + sRC} U_1(s)$$

This can be written as

$$U_2(s) = F(s)U_1(s) \qquad\qquad\qquad\qquad \text{(F-23)}$$

where $F(s)$ is the transfer function of the network. If $F(s)$ is known, the transient response of the network on any input signal $u_1(t)$ may be calculated.

In our example, $u_1(t)$ is a single square-wave pulse whose Laplace transform was previously determined in Sec. F-4-4. For $U_1(s)$ we can therefore write

$$U_1(s) = u_0 \frac{1 - e^{-s\tau}}{s}$$

where $\tau$ is the duration of the pulse and $u_0$ is its amplitude. Then $U_2(s)$ becomes

$$U_2(s) = u_0 \frac{1 - e^{-s\tau}}{s + 1/T} \tag{F-24$a$}$$

where $T = RC$.

To transform this expression back into the time domain, we decompose it as follows:

$$U_2(s) = u_0 \frac{1}{s + 1/T} - u_0 \frac{e^{-s\tau}}{s + 1/T} \tag{F-24$b$}$$

From the table of Laplace transforms (Table F-1), the signal corresponding to the first term of Eq. (F-24$b$) is

$$u_{21}(t) = u_0\, e^{-t/T} \qquad t \geq 0 \tag{F-25$a$}$$

The second term in Eq. (F-24$b$) corresponds to a decaying exponential function delayed by $\tau$,

$$u_{22}(t) = -u_0\, e^{-(t-\tau)/T} \tag{F-25$b$}$$

Because the signal $u_{21}(t)$ is zero for $t < 0$, the *delayed* signal $u_{22}(t)$ is defined only for $t \geq \tau$, but is zero for $t < 0$. Therefore, for the combined output signal $u_2(t)$ of the differentiator we obtain

$$u_e(t) = \begin{cases} 0 & t < 0 \\ u_0\, e^{-t/T} & 0 \leq t \leq \tau \\ u_0\, [e^{-t/T} - e^{-(t-\tau)/T}] & t > \tau \end{cases}$$

The waveform of $u_2(t)$ is plotted in Fig. F-14$c$.

## F-7 CLOSING THE GAP BETWEEN THE TIME DOMAIN AND THE COMPLEX-FREQUENCY DOMAIN

As demonstrated in the previous section, the Laplace transforms of input and output signals of any linear electric network are related by the transfer function $F(s)$,

$$U_2(s) = U_1(s)F(s) \tag{F-23}$$

Equation (F-23) enables us to determine the transient response of the network for any input signal $u_1(t)$.

Assume now that the network is excited by a delta function, $u_1(t) = \delta(t)$.

The transient response of the network on a delta function $\delta(t)$ will be hereafter denoted by $u_2(t) = h(t)$. As shown in Sec. F-4-5, the Laplace transform of the delta function is

$$\mathcal{L}\{\delta(t)\} = 1$$

Introducing this expression into Eq. (F-23) we obtain

$$U_2(s) = F(s)$$

From this expression we learn that the transfer function $F(s)$ of an electric network is the Laplace transform of the transient response $h(t)$ on a delta function

$$F(s) = \mathcal{L}\{h(t)\} \qquad\qquad\qquad (F\text{-}26)$$

A practical example will clarify the correspondence given by Eq. (F-26).

### Numerical Example

A passive $RC$ low-pass filter (commonly called an $RC$ integrator), as shown in Fig. F-15, is excited by a delta function

$$u_1(t) = \delta(t)$$

What is the transient response $u_2(t)$ on this input signal?

### Solution

Applying Ohm's law to the resistor and capacitor [Eqs. (F-21a) and (F-22b), we obtain for the transfer function

$$F(s) = \frac{U_2(s)}{U_1(s)} = \frac{1}{1 + sT}$$

where $T = RC$. According to Eq. (F-26), the transient response $u_2(t)$ of the integrator on a delta function applied to its input is

$$u_2(t) = \mathcal{L}^{-1}\{F(s)\} = \mathcal{L}^{-1}\left\{\frac{1}{1 + sT}\right\}$$

Using the table of Laplace transforms (Table F-1) we obtain

$$u_2(t) = h(t) = \frac{1}{T}\,e^{-t/T}$$

This is the impulse response plotted on the right-hand side of Fig. F-15.

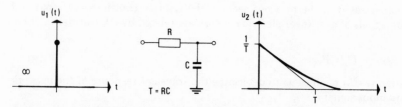

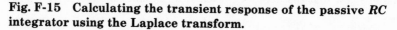

**Fig. F-15  Calculating the transient response of the passive $RC$ integrator using the Laplace transform.**

## F-8 NETWORKS WITH NONZERO STORED ENERGY AT $t = 0$

We shall now apply the Laplace transform to electric networks having either an inductor in which a nonzero current $i(0)$ flows at $t = 0$ or a capacitor on which there is a nonzero voltage $u(0)$ at $t = 0$. In both cases nonzero initial energy is stored in the network at $t = 0$. Let us again calculate the transient response of the $RC$ integrator in Fig. F-15 on a delta function, assuming now that the initial voltage across the capacitor is $u(0) \neq 0$.

Two equations in the time domain can be written for the $RC$ integrator:

$$u_1(t) = i(t)R + u_2(t)$$

$$u_2(t) = \frac{1}{C} \int i \, dt$$

Applying the Laplace transform to these equations [refer to Eqs. (F-21)], we obtain

$$U_1(s) = I(s)R + U_2(s)$$

$$U_2(s) = \frac{1}{C} \left[ \frac{I(s)}{s} + \frac{i^{(-1)}(0)}{s} \right]$$

We can eliminate $I(s)$ from the first of these equations. We then get

$$U_2(s) = \frac{1}{C} \left[ \frac{U_1(s) - U_2(s)}{Rs} + \frac{i^{(-1)}(0)}{s} \right]$$

After some manipulation, we have

$$U_2(s) = U_1(s) \frac{1}{1 + sT} + \frac{i^{(-1)}(0)}{C} \frac{T}{1 + sT}$$

where $T = RC$ and $i^{(-1)}(0)$ is the integral of current which flowed into the capacitor in the time interval $-\infty < t < 0$, that is, $i^{(-1)}(0)$ is the initial *charge* stored in the capacitor at $t = 0$. Hence $i^{(-1)}(0)/C$ is simply the initial voltage $u(0)$ across the capacitor. Consequently we obtain

$$U_2(s) = U_1(s) \frac{1}{1 + sT} + u(0) \frac{T}{1 + sT}$$

Because $u_1(t)$ has been assumed to be a delta function, $U_1(s) = 1$, and we have

$$U_2(s) = \frac{1}{1 + sT} + u(0) \frac{T}{1 + sT}$$

Transforming this equation back into the time domain (see Table F-1), we obtain for $u_2(t)$

$$u_2(t) = \frac{1}{T} e^{-t/T} + u(0) \, e^{-t/T}$$

Note that the first term is identical with the response of the *RC* integrator obtained for zero initial voltage across the capacitor. The second term is due to the initial charge stored in the capacitor.

## F-9 ANALYZING DYNAMIC PERFORMANCE BY THE POLE-ZERO PLOT

The transfer function of any linear network built from lumped elements such as resistors, inductors, capacitors, and amplifiers is given by a polynomial fraction in $s$,

$$F(s) = \frac{a_n s^n + a_{n-1} s^{n-1} + \cdots + a_1 s + a_0}{b_m s^m + b_{m-1} s^{m-1} + \cdots + b_1 s + b_2} \qquad m \geq n \qquad \text{(F-27a)}$$

This transfer function can also be written in the factored form:

$$F(s) = \frac{(s - \alpha_1)(s - \alpha_2) \cdots (s - \alpha_n)}{(s - \beta_1)(s - \beta_2) \cdots (s - \beta_m)} \qquad \text{(F-27b)}$$

Here the $\alpha_i$ values are the zeros and the $\beta_i$ values the poles of the transfer function.

The $\alpha_i$ and $\beta_i$ values can be either real or complex. For example, a real value of $\beta_i$ corresponds to a pole located on the real axis ($\sigma$ axis) in the complex $s$ plane (refer to Fig. F-16). If a complex pole is located at $\beta_i = A + jB$, another conjugate complex pole will exist at $\beta_i^* = A - jB$, where the asterisk denotes the conjugate value of $\beta_i$. Hence complex poles always exist as pairs of conjugate complex poles.

We will now see that the transient response of a network is very easily found if the locations of the poles and zeros of the network transfer function $F(s)$ are known. To transform Eq. (F-27b) back into the time domain, it is most convenient to decompose this expression into partial fractions;

$$F(s) = \underbrace{\frac{R_1}{s - \beta_1} + \frac{R_2}{s - \beta_2} + \cdots}_{\substack{\text{partial fractions generated} \\ \text{by single real poles}}}$$

$$\underbrace{+ \underbrace{\frac{R_i}{s - (A_i + jB_i)} + \frac{R_i^*}{s - (A_i - jB_i)}}_{\substack{\text{2 partial fractions generated by} \\ \text{one pair of conjugate complex poles}}} + \cdots}_{\substack{\text{partial fractions generated by} \\ \text{all pairs of conjugate complex poles}}} \qquad \text{(F-27c)}$$

The terms $R_1, R_2, \ldots$ are constants and are called *residues*.[20] In Eq. (F-27c) we separated two groups of partial fractions, those emanating from the single real poles, and those emanating from the pairs of conjugate complex poles.

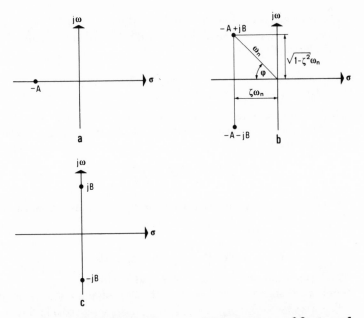

**Fig. F-16  Calculating the transient response of first- and second-order systems from the pole positions. (*a*) First-order system, pole on the negative $\sigma$ axis. (*b*) Second-order system, complex-conjugate pole pair in the negative half-plane (damped oscillation). (*c*) Second-order system, poles on the imaginary axis (undamped oscillation).**

Let us first look at the transient response $f(t)$ due to the real poles of $F(s)$. The inverse Laplace transform of the term

$$F(s) = \frac{R_i}{s - \alpha_i}$$

is

$$f(t) = R_i \exp(\alpha_i t)$$

Hence the contribution of the single real poles to the signal $f(t)$ is

$$f(t)_{\text{single poles}} = R_1 \exp(\alpha_1 t) + R_2 \exp(\alpha_2 t) + \cdots \tag{F-28}$$

Note that the residues of the partial fractions due to single real poles are always real numbers. [If they are not, the signal $f(t)$ would become complex, which is physically impossible.]

The next portion of the transient response $f(t)$ is contributed by the complex pole pairs. For simplicity we isolate one complex pole pair:

$$F(s)_{\text{pole pair}} = \frac{R_i}{s - (A_i + jB_i)} + \frac{R_i^*}{s - (A_i - jB_i)} \tag{F-29a}$$

Here the residues $R_i$ and $R_i^*$ must not necessarily be real, but can be complex. Moreover, if $R_i$ is a complex number, $R_i^*$ is its conjugate value, i.e. in the most general case we have

$$R_i = a + jb$$

$$R_i^* = a - jb$$

where $a$ and $b$ are real constants.

This is easily proved by combining the two terms in Eq. (F-29a) over a common denominator,

$$F(s)_{\text{pole pair}} = \frac{s(R_i + R_i^*) - A(R_i + R_i^*) + jB(R_i - R_i^*)}{s^2 - 2As + (A^2 + B^2)} \qquad \text{(F-29b)}$$

All individual coefficients in the numerator of Eq. (F-29b) must be real, that is,

$$R_i + R_i^* \rightarrow \text{real}$$

$$j(R_i - R_i^*) \rightarrow \text{real or } R_i - R_i^* \rightarrow \text{imaginary}$$

These conditions are met only if $R_i$ and $R_i^*$ form a conjugate complex pair of numbers, as stated above. Equation (F-29a) can therefore be rewritten as

$$F(s)_{\text{pole pair}} = \frac{a + jb}{s - (A + jB)} + \frac{a - jb}{s - (A - jB)} \qquad \text{(F-29c)}$$

Referring to Table F-1, we recall that the inverse Laplace transform of $1/(s - \alpha)$ is $f(t) = e^{\alpha t}$. This also applies for complex $\alpha$. Applying this to Eq. (F-29c), we obtain

$$f(t)_{\text{pole pair}} = a[e^{AT}e^{jBt} + e^{At}e^{-jBt}] + jb[e^{At}e^{jBt} - e^{At}e^{-jBt}] \qquad \text{(F-29d)}$$

Using the well-known Euler theorem,

$$\sin x = \frac{e^{jx} - e^{-jx}}{2j}$$

$$\cos x = \frac{e^{jx} + e^{-jx}}{2}$$

we can write

$$f(t)_{\text{pole pair}} = 2a\, e^{At} \cos Bt - 2b\, e^{At} \sin Bt \qquad \text{(F-29e)}$$

This can be brought into the more general form

$$f(t)_{\text{pole pair}} = C\, e^{At} \cos (Bt + \phi) \qquad \text{(F-29f)}$$

where $C = 2 \sqrt{a^2 + b^2}$

$$\phi = -\tan^{-1} \frac{b}{a}$$

The results are summarized in Table F-2.

**Table F-2. Laplace Transforms of Generalized First- and Second-Order Systems**

| $F(s)$ | $f(t)$ |
|---|---|
| $\dfrac{R_i}{s - \alpha_i}$ | $\rightarrow R_i\, e^{\alpha_i t}$ |
| $\dfrac{R_i}{s - (A + jB)} + \dfrac{R_i^*}{2 - (A - jB)}$ | $\rightarrow C\, e^{At} \cos{(Bt + \phi)}$ |

These results allow a simple interpretation of the term complex frequency, introduced earlier, which will be discussed in the next section.

## F-10 A SIMPLE PHYSICAL INTERPRETATION OF "COMPLEX FREQUENCY"

Refer again to the pole-zero plot in Fig. F-16. A single pole located on the negative $\sigma$ axis ($\alpha < 0$) gives rise to the decaying exponential function (see Table F-2)

$$f(t) \sim e^{\alpha t} \qquad \alpha < 0$$

This is an exponential function having a real exponenent; hence $\alpha$ is called a *real frequency*.

A complex pole pair located on the imaginary axis [$A = 0$ in Eq. F-29$f$] gives rise to an undamped oscillation:

$$f(t) \sim \cos Bt$$

This is an exponential function having a purely imaginary exponent; hence $B$ is called an *imaginary frequency*.

A pole pair located in the left half of the $s$ plane, with $A < 0$ and $B \neq 0$, gives rise to a damped oscillation:

$$f(t) = e^{At} \cos{(Bt + \phi)}$$

This is an exponential function having a complex exponent; hence we speak of an oscillation with a *complex frequency*.

It has proved useful to write partial fractions generated by complex conjugate pole pairs in the so-called normalized form. When doing so, we introduce the substitution [refer to Eq. (F-29$c$)];

$$F(s)_{\text{pole pair}} = \frac{a + jb}{s - (A + jB)} + \frac{a - jb}{s - (A - jB)} \tag{F-29$g$}$$

$$\rightarrow \frac{a + jb}{s - (-\zeta\omega_n + j\sqrt{1 - \zeta^2}\,\omega_n)} + \frac{a - jb}{s - (-\zeta\,\omega_n - j\sqrt{1 - \zeta^2}\,\omega_n)}$$

Comparing the coefficients on both sides of the arrow, we obtain the equalities

$$A \equiv -\zeta \omega_n \qquad \text{(F-30)}$$
$$B \equiv \sqrt{1 - \zeta^2}\, \omega_n$$

where $\zeta$ and $\omega_n$ are the damping factor and the natural frequency, respectively.

If a transfer function $F(s)$ has a conjugate complex pole pair located at $A \pm jB$ in the complex $s$ plane, this pole pair will generate a transient response $f(t)$ of the form

$$f(t) = \exp(-\zeta\, \omega_n t)\, \cos\left(\sqrt{1 - \zeta^2}\, \omega_n t + \phi\right) \qquad \text{(F-31)}$$

The time constant of the decaying exponential function in Eq. (F-31) is given by $1/(\zeta\, \omega_n)$; the frequency of the damped oscillation is given by $\omega_{res} = \sqrt{1 - \zeta^2}\, \omega_n$. If the complex pole pair is plotted in the complex $s$ plane, the distance of the poles from the origin is seen to be exactly $\omega_n$.

# REFERENCES

1. Gardner, Floyd M.: *Phaselock Techniques,* 2d ed., copyright © John Wiley and Sons, New York, 1979.
2. Richman, D.: "Color Carrier Reference Phase Synchronization Accuracy in NTSC Color Television," *Proc. IRE,* vol. 42, January 1954, pp. 106–133.
3. Izawa, K.: *Introduction to Automatic Control,* Elsevier, New York, 1963.
4. Chestnut, H., and R. W. Mayer: *Servomechanisms and Regulating System Design,* vol. 1, 2d ed., Wiley, New York, 1961.
5. Viterbi, Andrew J.: "Acquisition and Tracking Behaviour of Phase-Locked Loops," Jet Propulsion Laboratory, External Publication 673, July 14, 1959.
6. Frazier, J. P., and J. Page: "Phase Lock Loop Frequency Acquisition Study," *IRE Trans. Space Electron, Telem.,* Vol. SET-8, September 1962, pp. 210–227.
7. Sanneman, R. W., and J. R. Rowbotham: "Unlock Characteristic of the Optimum Type II Phase-Locked Loop, *IEEE Trans. Aerosp. Navig. Electron.,* Vol. ANE-11, March 1964, pp. 15–24.
8. *Phase-Locked Loop Data Book,* 2d ed., Motorola Semiconductor Products Inc., Phoenix, Arizona, August 1973.
9. Reed, L. J., and Ron J. Treadway: "Test your PLL IQ," *EDN,* Dec. 20, 1974.
10. Gardner, Floyd M.: "Charge-Pump Phase-Lock Loops," *IEEE Trans. Commun.,* vol. COM-28, November 1980.
11. Lindsey, William C., and Chak Ming Chie: "A Survey of Digital Phase-Locked Loops," *Proc. IEEE,* Vol. 69, April 1981.
12. Larimore, W. E.: "Synthesis of Digital Phase-Locked Loops," in 1968 EASCON Rec., October 1968, pp. 14–20.
13. Larimore, W. E. "Design and Performance of a Second-Order Digital Phase-Locked Loop," presented at the Symposium on Computer Processing in Communications (Polytech. Inst. of Brooklyn, New York), April 8–10, 1969, pp. 343–357.
14. Gill, G. S. and S. C. Gupta: "First-Order Discrete Phase-Locked Loop with Applications to Demodulation of Angle-Modulated Carrier," *IEEE Trans. Commun. Technol.,* vol. CT-20, June 1972, pp. 454–462.
15. Gill, G. S., and S. C. Gupta, "On Higher-Order Discrete Phase-Locked Loops," *IEEE Trans. Aerosp. Electron. Syst.,* vol. AES-8, September 1972, pp. 615–623.
16. Langston J. Leland, "$\mu$C Chip Implements High-Speed Modems Digitally," *Electron. Des.,* June 24, 1982.
17. Best R. (Ed.): *Handbuch der analogen und digitalen Filterungstechnik* (in German), AT-Verlag, Aarau, Switzerland, 1982.
18. Greer, W. T., Jr., and Bill Kean: "Digital Phase-Locked Loops Move into Analog Territory," *Electron. Des.,* March 31, 1982.
19. Stofka, Marian: "Digital-Only PLL Exhibits No Overshoot," *EDN,* May 26, 1982.
20. Tou, Julius T.: *Digital and Sampled-Data Control Systems,* McGraw-Hill, New York, 1959.

21. Temes, Gabor C., and Sanjit K. Mitra: *Modern Filter Theory and Design,* Wiley, New York, 1973.
22. Troha, Donald G., and James D. Gallia: "Digital Phase-Locked Loop Design Using SN 54/74 LS 297," Application Note AN 3216, Texas Instruments Inc., Dallas, Texas.
23. Isbell, T. D., and D. S. Mishler: "LM 1800 Phase-Locked Loop FM Stereo Demodulator, Application Note AN-81, National Semiconductor Corp.
24. Briant, James M.: "SL 650 & SL 651 Applications," Application Note, Plessey Semiconductors, Wiltshire SN2 2QW, UK.
25. *Analog-Digital Conversion Handbook,* Analog Devices Inc., Norwood, Massachusetts, 1972.
26. Bently, W. E., and S. G. Varsos: "Squeeze More Data onto Mag Tape by Use of Delay-Modulation Encoding and Decoding," *Electron. Des.,* October 11, 1975.
27. *Reference Data for Radio Engineers,* International Telephone and Telegraph Corp.
28. Lyon, D.: "How to Evaluate a High-Speed Modem," *Telecommunications,* vol. 9, October 1975.
29. Mollinga, Th.: "AM Receivers with PLL Techniques," *EDN,* February 20, 1975.
30. Kuo, F. F.: *Network Analysis and Synthesis,* Wiley, New York, 1962.
31. Wong, Y. J., and W. E. Ott: *Function Circuits, Design and Applications,* McGraw-Hill, New York, 1976.
32. "Data Sheet of the Single Tone Switch FX-101L," Consumer Microcircuits Limited, Witham, Essex CM8 3TD, UK.
33. "Data Sheet of the 5-Tone Encoder/Decoder (Transceiver) FX-407," Consumer Microcircuits Limited, Witham, Essex CM8 3TD, UK.
34. Graeme, Jerald G.: *Applications of Operational Amplifiers,* McGraw-Hill, New York, 1973.
35. "Signetics Digital, Linear MOS Applications," Signetics Corp., pp. 6–59, 1974.
36. "New Synthesizer Circuits from Plessey," Publ. P.S. 1736, Plessey Semiconductors, Wiltshire, SN2 6BA, UK, June 1981.
37. "New Phase-Locked Loops Have Advantages as Frequency to Voltage Converters (and More)," Application Note AN 210, National Semiconductor Corp., April 1979.
38. Hatchett, John: "Frequency-Divider Systems Increase Flexibility, Save Parts," *EDN,* June 9, 1982.
39. Rohde, Ulrich: "Low-Noise Frequency Synthesizers Using Fractional $N$ Phase-Locked Loops," *r.f. design,* January/February 1981.
40. Graeme Jerald G.: *Designing with Operational Amplifiers,* McGraw-Hill, New York, 1977.
41. Counts, Lew, and Scott Wurcer: "Instrumentation Amplifier Nears Input Noise Flow," *Electron. Des.,* June 10, 1982.
42. Moore, A. W. "Phase-Locked Loops for Motor Speed Control," *IEEE Spectrum,* April 1973, pp. 61–67.
43. Smithgall, D. H.: "A Phase-Locked Loop Motor Control System," *IEEE Trans. Ind. Electron. Contr Instrum.,* vol. IECI-22, November 1975, pp. 487–490.
44. N. K. Sinha, "Speed Control of a DC Servo Motor Using Phase-Locked Loop," *IEEE Trans. Ind. Electron. Contr. Instrum.,* vol. IECI-23, February 1976, pp. 22–26.
45. Le-Huy, Hoang, and O. L. Mercier: "A Synchronous DC Motor Speed Control System," *Proc. IEEE,* March 1976, pp. 394–395.

46. Harashima, F., H. Naitoh, and H. Taoka: "A Microprocessor-Based PLL Speed Control System Converter-Fed Synchronous Motor," *IEEE Trans. Ind. Electron. Contr. Instrum.*, vol. IECI-27, August 1980.
47. Feller, D. W.: "Design CMOS Commutative Filters," *Electron. Des.*, November 8, 1974.
48. Gasquett, H.: "A Monolithic Integrated FM Stereo Decoder System," Application Note AN-432 A, Motorola Semiconductor Products Inc., Phoenix, Arizona.
49. "New Techniques for Analyzing Phase Lock Loops," Application Note 164-3, Hewlett-Packard, June 1975.
50. Viterbi, Andrew J.: *Principles of Coherent Communication*, McGraw-Hill, New York, 1966.

# INDEX

AC component, 7
Acceleration error, 51, 53
Accumulator (ACCU), 228–229
Acquisition in time domain, 25–35
Acquisition performance, computer simulation of, 45
Acquisition process, 15
Active loop filter, 114–115
Active-low signals, 12
ADC (*see* Analog-to-digital converter)
Addition theorem, Laplace transform, 313
Admittance, measurement of, 235–237
ADPLL (*see* All-digital PLL)
AF (audio-frequency) signal, 155
Aliasing effects, 102
All-digital loop filters, 96–102
All-digital PDs, 89–96
All-digital PLL (ADPLL), 69, 86
  functional blocks in, 88–89
  implementations of, 105–110
AM (*see* Amplitude modulation)
Amplifier, instrumentation, 237
Amplitude, 304–305
Amplitude distortion, 177
Amplitude modulation (AM), 168–171
  modulation and demodulation, 178–195
  quadrature (QAM), 176–177
  receiver, 153
    experimental circuit of, 190, 191
    waveforms of, 196
Amplitude spectrum, 305
Analog multiplier, 7, 181
Analog signals, 165
  intermediate (*see* Intermediate analog signals)
Analog-to-digital converter (ADC), 92
  using VCO, 254–255
Angular frequency, 16
  of output signal, 2, 3
  rate of change of, 16
  of reference signal, 1–3

Angular frequency error, 80
Antiphase, 187
ASCII code, 165
Audio-frequency (AF) signal, 155
Average phase, 263

Bandwidth:
  noise, 58
  of synchronous filter, 251
  3-dB, measurement of, 278–281
Base modulator, 178, 180, 181
BCD (binary-coded decimal) codes, 165–166
Bessel function, 172
Binary-coded decimal (BCD) codes, 165–166
Binary codes, 165
Binary counter, 88
Binary data signals, modulation of, 174–177
Binary keying input, 146
Binary signal, 86, 165
Binary-valued output signals, 96
Biphase formats, 166–168
Bit, most significant (MSB), 265, 267
Bode diagram, 308
  of error-transfer function, 24
  of first-order loop filters, 12
  of motor-speed control system, 242–24^
  of phase-transfer function, 23
  of tracking filter, 162–163
  of transfer functions, 23–24
BORROW pulses, 99
Buffer amplifier, 195

Capacitor, impedance of, 321
Capture range, 35
Carrier amplitude, 169
CARRY pulses, 99
CB transceivers, 217
CCO (current-controlled oscillator), 1

335

# ABOUT THE AUTHOR

Roland E. Best, Ph.D., is a world-renowned authority on phase-locked loops, circuit design, and microprocessor applications. His articles on these subjects have been published by such journals as *Der Elektroniker*, *Electronic Design*, *EDN*, *Instrumentation Technology*, and *Electronic Products*. This book is largely based on the author's own highly acclaimed book on phase-locked loops, first published in German in 1976 and now in its third edition.

Educated at the prestigious Swiss Federal Institute of Technology, Dr. Best worked at IBM for several years before joining the senior staff at Sandoz AG in Basel, Switzerland.